MEDICAL
INTELLIGENCE
UNIT

INTRACELLULAR PARASITISM

OF MICROORGANISMS

Springer-Verlag
Berlin Heidelberg GmbH

R.G. LANDES COMPANY
AUSTIN

MEDICAL INTELLIGENCE UNIT
INTRACELLULAR PARASITISM OF MICROORGANISMS

R.G. LANDES COMPANY
Austin, Texas, U.S.A.

International Copyright © 1996 Springer-Verlag Berlin Heidelberg
Originally published by Springer-Verlag Heidelberg, Germany in 1996
Softcover reprint of the hardcover 1st edition 1996

 Springer

International ISBN 978-3-662-22049-8

While the authors, editors and publisher believe that drug selection and dosage and the specifications and usage of equipment and devices, as set forth in this book, are in accord with current recommendations and practice at the time of publication, they make no warranty, expressed or implied, with respect to material described in this book. In view of the ongoing research, equipment development, changes in governmental regulations and the rapid accumulation of information relating to the biomedical sciences, the reader is urged to carefully review and evaluate the information provided herein.

Library of Congress Cataloging-in-Publication Data
Belyi, Yuri F., 1960-
    Intracellular parasitism of microorganisms/ Yuri F. Belyi.
        p. cm. — (Medical intelligence unit)
    Includes bibliographical references and index.
    ISBN 978-3-662-22049-8   ISBN 978-3-662-22047-4 (eBook)
    DOI 10.1007/978-3-662-22047-4
    1. Host-bacteria relationships. 2. Cellular signal transduction. 3. Legionella. 4. Listeria. 5. Mycobacterium. I. Title. II. Series.
        [DNLM: 1. Bacteria--metabolism. 2. Bacteria--pathogenicity. 3. Eukaryotic Cells--parasitology. 4. Eukaryotic Cells--metabolism. 5. Signal Transduction. QW 52B4541 1996]
QR175.B45 1996
DNLM/DLC   96-18622
for Library of Congress   CIP

Publisher's Note

R.G. Landes Company publishes six book series: *Medical Intelligence Unit, Molecular Biology Intelligence Unit, Neuroscience Intelligence Unit, Tissue Engineering Intelligence Unit, Biotechnology Intelligence Unit and Environmental Intelligence Unit.* The authors of our books are acknowledged leaders in their fields and the topics are unique. Almost without exception, no other similar books exist on these topics.

Our goal is to publish books in important and rapidly changing areas of bioscience and environment for sophisticated researchers and clinicians. To achieve this goal, we have accelerated our publishing program to conform to the fast pace in which information grows in bioscience. Most of our books are published within 90 to 120 days of receipt of the manuscript. We would like to thank our readers for their continuing interest and welcome any comments or suggestions they may have for future books.

Deborah Muir Molsberry
Publications Director
R.G. Landes Company

CONTENTS

Interaction of bacteria with macroorganisms is a complex multifactorial process which starts with attempts by microbes to penetrate host barriers. Such attempt is normally confronted by diverse, more or less, specific mechanisms accomplished by specialized host tissues and directed to elimination of the invaders. Usually this antibacterial attack results in eradication of intruders. However certain microorganisms known as pathogenic bacteria are equipped to resist the primary bactericidal response of the target organism and are allowed to proliferate as parasites.

Among these microorganisms there is a group of facultative intracellular parasites which is distinguished by its capacity to use phagocytic cells in addition to extracellular space as a niche for replication. The typical examples of facultative intracellular parasites include *Legionella*, *Listeria* and *Mycobacterium* species (excluding probably only *M.leprae* which is an obligate intracellular parasite), replicating both in "professional" and "non-professional" phagocytes.

Since following phagocytosis, bacteria are opposed by potent eukaryotic equipment for killing and degradation of internalized organisms, facultative intracellular parasites during evolution obtained sophisticated mechanisms to avoid or decrease such bactericidal attack of phagocytes. Among these mechanisms are specialized composition of the cell envelope which helps bacteria to resist bacteriolytic activity of lysosomal contents, eleboration of enzymes which scavenge or degrade bactericidal eukaryotic products, synthesis of toxic substances which produce general damage to phagocytes, lysing of phagosomal membrane and intracytoplasmic proliferation etc. However, theoretically the most rational strategy from the point of view of the parasite could be not the degradation of already synthesized bactericidal agents, but interference with their synthesis and—first of all—alteration in the regulation of biosynthesis.

It is known that regulation of every metabolic event, including synthesis of bactericidal proteins, oxidative burst generation, acidification of phagosomes etc. is under the accurate control of a cell and is accomplished through signaling pathways. Correspondingly such signaling processes apparently could be one of the most advantageous targets for facultative intracellular parasites.

Indeed, the literature data and the data from author's laboratory illustrate the hypothesis that one of the ways by which intracellularly multiplying bacteria manipulate eukaryotic functions and thus influence the bactericidal response of phagocytes is by altering signaling cascades in eukaryotes. To achieve this goal bacteria are able to produce enzymes and other biologically active compounds which misdirect signaling pathways for the benefit of the parasites. Accordingly, this manuscript is an attempt of the author to look at the problem of intracellular parasitism of bacteria from the "signaling" point of view.

The book is composed of two main parts. The first one provides general information on signaling cascades in eukaryotic cells. Signal transduction in eukaryotic organisms is an area of modern biology which attracts intense attention and is one of the most quickly developing. Therefore the information present in the first sections of the manuscript tends to be (1) archaic and (2) imperfect in outlining the fine details of signaling mechanisms. However this manuscript does not pretend to be a handbook on signaling cascades and it is pretty well understood that every chapter describing mechanisms of signal transduction could be a theme for separate volumes. On the contrary, the idea is to give a general overview of the problem of regulation within a cell and to make an effort to present signal transduction as an undivided network of communicated processes linking specific signals with definite responses. Therefore the main purpose of this presentation of intracellular signaling "at a glance" is to create a background for discussion of bacteria-target cell interaction.

The second section describes *Legionella*, *Listeria* and *Mycobacterium* factors exercising regulatory activity toward the metabolism of host cells, essentially by affecting signal transduction in eukaryotes. The data on distinct products of the bacteria is discussed in relation to the biology of intracellular parasitism of these pathogenic microorganisms. Such an overlay of microbial pathogenesis and intracellular signaling is designed to help clarify the role of known bacterial molecules and to circumscribe possible aims and areas of future research.

Acknowledgments

I'd like to thank my colleagues from N.F. Gamaleya Research Institute, Moscow, Russia for help in research on *Legionella* and *Listeria* and Dr. Yuri Nedialkov, Michigan State University, for assistance in literature database searches. I wish to acknowledge Ronald G. Landes, M.D., and the staff of R.G. Landes Company for the opportunity to publish this book. Special thanks to Francine Daniel, Lynn O'Neill and Mary Kelly for persistence and cooperativity. I beg pardon of authors whose papers were not cited in this manuscript due to space limitations.

SECTION I

SIGNALING IN
EUKARYOTIC CELLS

RECEPTORS

The transduction of an extracellular signal inside the cell, and triggering of complex eukaryotic signaling machinery, in many instances begins with binding of a specific ligand to its specific plasma membrane receptor. Mechanisms for transduction of signals across the plasma membrane are various and determine structural diversity of different groups of receptors. They can be divided into two main groups. The first one includes receptors that possess intrinsic catalytic activity or ion-channel activity. These constructions can be considered as enzymes or ion channels with additional domains necessary for agonist binding. In many instances such receptors contain an accessory hydrophobic transmembrane domain which is responsible for proper orientation of the molecule and transduction of received signal across the plasma membrane. Binding of the agonist results in complex structural rearrangements of the receptor molecule, which resolve in alterations of receptor enzymatic activity or in changes of membrane permeability for certain ions. The second group contains receptors which are enzymatically non-active, but structurally linked with downstream signaling complexes. Such receptors are composed of sensory, transmembrane and specialized intracellular domains. The latter most often were shown to be implicated in transduction of an extracellular signal to coupled intracellular signal transducing molecule(s).

RECEPTOR PROTEIN KINASES

In a group of receptor protein kinases, agonist binding activates an intrinsic protein kinase catalytic domain with subsequent phosphorylation of cytoplasmic substrates. Receptor protein kinases display substrate specificity to phosphorylate either tyrosine or serine/threonine residues in their targets. Whereas receptor protein serine/threonine kinases remain relatively poorly studied, receptor protein tyrosine kinases (RPTKs) have been investigated in more detail.

To date, over 50 RPTK genes are described. They are subdivided into 14 main families on the basis of their catalytic domain sequences and on the overall structure.[387] Representatives of these families are epidermal growth factor (EGF), platelet-derived growth factor (PDGF), insulin, fibroblast growth factor (FGF), nerve growth factor (NGF), hepatocyte growth factor (HGF) receptors and some others. Knowledge about ligands as well as substrates for about half of the RPTKs is however lacking. Therefore the functions of many receptors remain unclear.

Typically RPTKs are composed of a large stretch of extracellular amino-terminal glycosylated domains necessary for agonist binding, a membrane spanning region, spacer region, a protein tyrosine kinase catalytic domain and an intracellular carboxyterminal tail (Fig. 1). The catalytic domain of RPTK is the most conserved domain of these receptors and catalyzes the transfer of the γ-phosphate of ATP to tyrosine residues on a substrate protein and/or onto the receptor molecules themselves in an autophosphorylation reaction.[182,292]

Signal transduction starts when the ligand binds the receptor. This process triggers dimerization of RPTKs.[164,386] Dimerization draws together two catalytic domains which then leads to conformational changes with subsequent phosphorylation of certain tyrosine residues in the cytoplasmic domains. Such autophosphorylation is commonly seen on conserved tyrosine residues within the kinase domains. These residues appear to be allosteric sites that regulate the specific activity of the receptor kinase. Not all receptors are however regulated by phosphorylation inside the kinase domain; e.g., in the EGF receptor the conserved tyrosine residue in the kinase domain appears not to be autophosphorylated.[169]

The other class of autophosphorylation sites are normally localized outside the kinase domains and serve the important func-

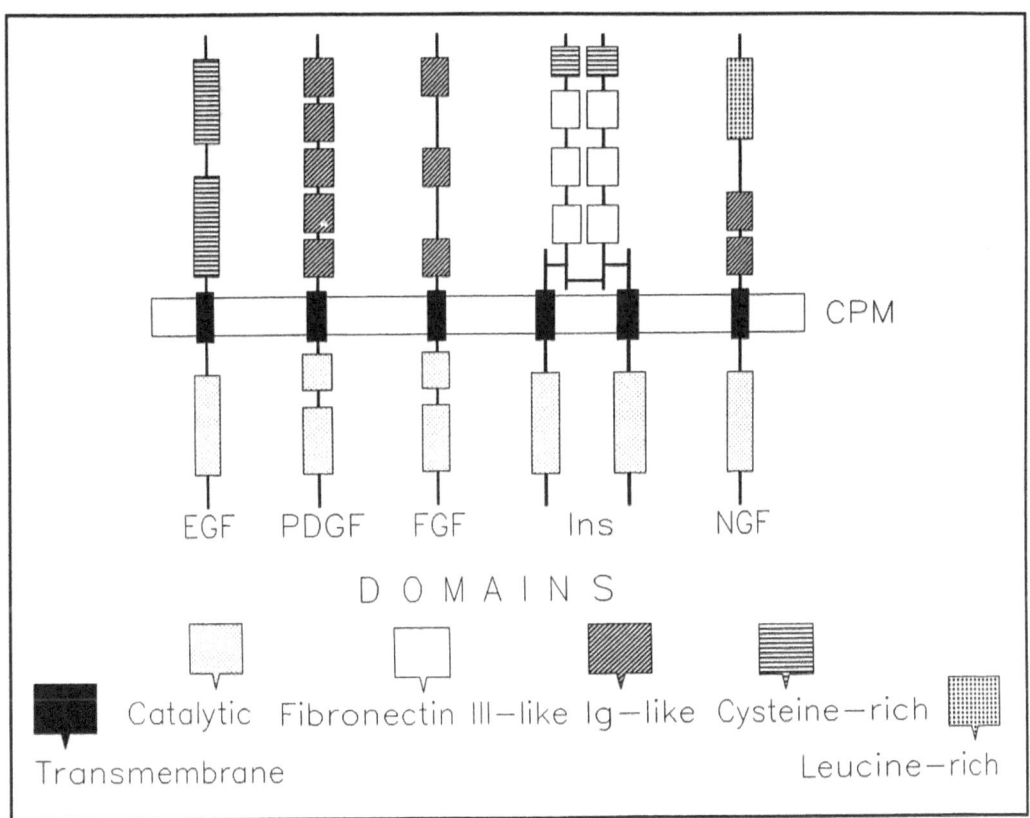

Fig. 1. Structural organization of receptor protein tyrosine kinases. EGF, epidermal growth factor receptor; PDGF, platelet-derived growth factor receptor; FGF, fibroblast growth factor receptor; Ins, insulin receptor; NGF, nerve growth factor receptor; CPM, cytoplasmic membrane.

tion of creating docking sites for downstream signal transduction molecules. These molecules, to be engaged by the receptor, in addition to domains responsible for enzymatic activity (e.g., phospholipid hydrolysis, protein phosphorylation, phosphoprotein dephosphorylation etc.) should contain one or several copies of a protein module of approximately 100 amino acid residues in length, the SH2 (Src homology 2) domain. SH2 motifs recognize phosphotyrosine-containing sites and are responsible for association of such SH2-containing proteins with autophosphorylated receptors.

At least three classes of phospholipases (PLCγ, PLA$_2$ and PLD), Ras guanine nucleotide dissociation stimulators, c-Src protein kinases and phosphatidylinositol 3-kinases can be recruited by RPTKs via their SH2 domains. In addition, at least two SH2 domain-containing protein tyrosine phosphatases, SH-PTP1 and SH-PTP2, activated through binding to and phosphorylation by tyrosine-phosphorylated receptors, have been described. SH-PTP1 was shown to participate in regulation of receptor activity since it dephosphorylated activated RPTKs and thus might be involved in its down regulation.[387]

In addition to these enzymes, there are RPTK-recruited proteins that lack an obvious catalytic domain. These proteins commonly contain SH2 and SH3 ("Src homology 3") domains or may bind proteins containing such domains and have become known as "adapter proteins" and "docking proteins" respectively. They are believed to serve as intermediates between RPTKs and downstream signaling molecules. Examples of this category are Grb2 protein, which upon binding to a phosphorylated RPTK leads to activation of the Ras signaling pathway, and IRS-1, which can bind multiple SH2 proteins following phosphorylation. These examples illustrate that not all the proteins that associate functionally with RPTKs need to be substrates, i.e., need to be phosphorylated. A case in point is Grb2, which is not phosphorylated on tyrosine residues in stimulated cells, yet binds functionally to activated RPTKs. Such proteins have been proposed to be designated as "targets" rather than "substrates".[387]

The binding of a cytoplasmic SH2-containing signaling protein to an activated receptor could modulate activity of the former in several different ways. Firstly, receptor autophosphorylation may provide a means of recruiting SH2-containing proteins to the membrane. Because the substrates of these proteins in many examples are located at the membrane (e.g., phospholipids in the case of phospholipase Cγ or Ras-GDP in the case of Ras guanine nucleotide dissociation stimulators), this inducible association with the membrane may represent a mechanism of switching on such recruited enzymes. Secondly, the physical complex formed between a receptor tyrosine kinase and SH2-containing signaling protein may identify the latter as a preferential substrate for RPTK and might therefore contribute to its activation by phosphorylation. Thirdly, the mere binding of the SH2 domains of a signaling protein to phosphotyrosine-containing sites can produce conformational change necessary for modification of the activity of the associated catalytic domain.[292]

RECEPTOR PROTEIN PHOSPHATASES

Among the receptor protein phosphatases the most studied are receptors which dephosphorylate proteins containing phosphotyrosine residues—receptor protein tyrosine phosphatases (RPTPs). The representatives of this family are similar in overall structural composition to receptor protein tyrosine kinases and typically have an amino-terminal glycosylated extracellular region, a single membrane-spanning region, and a cytoplasmic region with catalytic domain(s) (Fig. 2). Approximately 20 receptor-type protein tyrosine phosphatases have been cloned, not including the various homologs from different species.[71,182,380,396]

Based on the structural variations of the extracellular region, RPTPs have been subdivided into four groups.[119] According to this classification, the type I protein tyrosine phosphatases have a highly glycosylated extracellular region that is not homologous to any other known proteins. CD45 is a representative example of this group. The type II RPTPs have extracellular regions that are composed of varying numbers of Ig-like domains at the N-termini and fibronectin type III-like repeats. The type III RPTPs consist of only tandem fibronectin type III-like repeats in the extracellular part. The type IV RPTPs are defined by short extracellular regions. In addition to these four main subgroups, another type V group has been proposed which is characterized by the presence of a region with striking sequence similarities to the zinc-containing carbonic anhydrase at the extreme N-terminus.

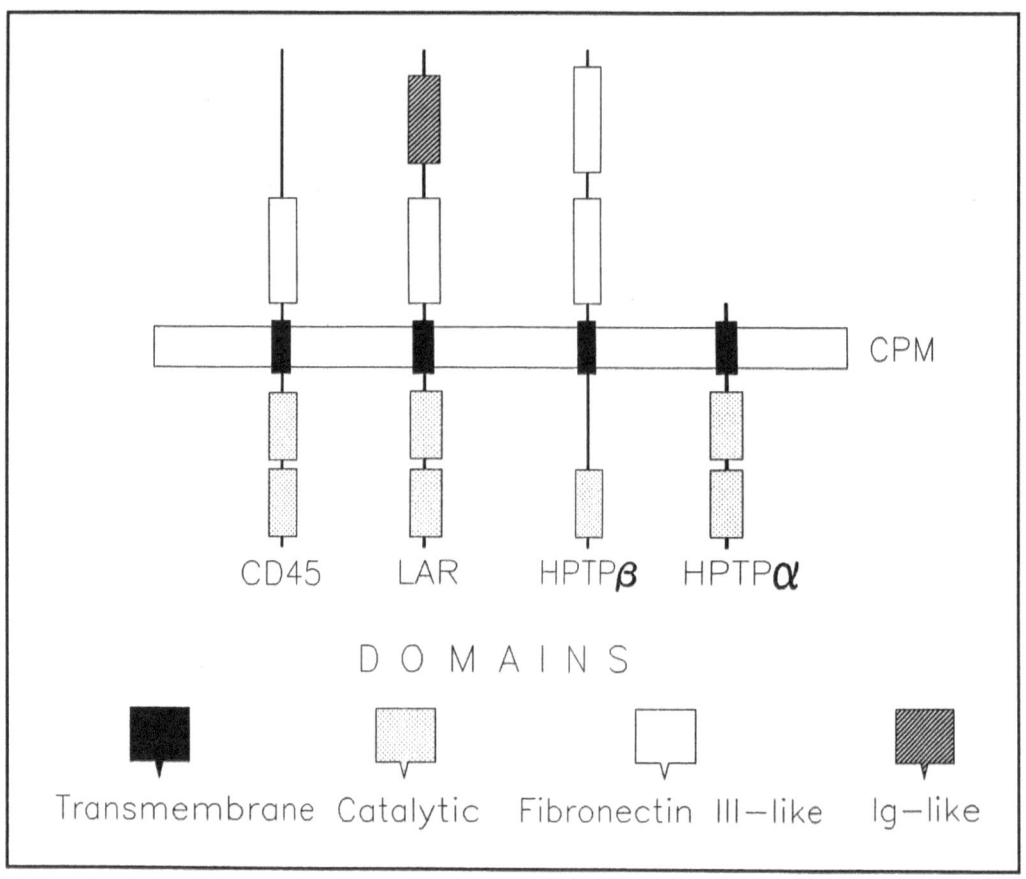

Fig. 2. Structural organization of receptor protein tyrosine phosphatases. LAR, leukocyte common antigen-related phosphatase; HPTP, human protein tyrosine phosphatase; CPM, cytoplasmic membrane.

One of the first identified and best studied RPTPs is the CD45 receptor, also named leukocyte common antigen (LCA), Ly-5, T200 or B220. It is a transmembrane glycoprotein expressed on nucleated cells of hematopoietic origin and was initially characterized as an abundantly expressed polymorphic lymphocyte-specific antigen.[374,405]

CD45 is characterized (as are a majority of RPTPs) by the presence of two phosphatase domains in the cytoplasmic region. The role of each domain in mediating phosphatase activity is not currently clear. For example, individual point mutations of the conserved cysteine residue, found in all phosphatase domains and involved in the phosphotransfer reaction, revealed that only the membrane proximal domain of CD45 had in vitro tyrosine phosphatase activity against a variety of artificial substrates.[151,305] Although the distal phosphatase domain did not appear to have catalytic activity, it was required for optimal function of the membrane proximal phosphatase domain, since deletion of the distal domain resulted in a loss of catalytic activity of the proximal domain.[189,361] It was concluded therefore that the second tyrosine phosphatase domain of this receptor-type phosphatase may regulate the activity of the first domain. Another possibility is that a "nonfunctional" protein tyrosine phosphatase domain could function in the manner of a SH2 domain, and induce the noncovalent association of the RPTP with tyrosine-phosphorylated protein.

Since no ligands have been conclusively described for CD45, the exact mechanism for regulation of phosphatase activity in this RPTP remains unknown. Several propositions have been suggested. Given the structural similarities of CD45 to growth factor receptors, one potential means of regulation could be through binding of ligand to the extracellular domain. It may result in dimerization and activation of CD45. Due to the fact that CD45 exhibits high intrinsic catalytic activity in vitro, it is plausible that the phosphatase activity is regulated negatively by ligand binding to the extracellular domain.

The next possible mode of regulation could be by phosphorylation. In vitro phosphorylation of RPTP has been achieved with the tyrosine kinases Abl and Src, the serine/threonine kinases casein kinase 2, protein kinase C and glycogen synthase kinase 3. However no change in phosphatase activity was detected. In other experiments, activity of CD45 has been shown to be transiently increased by the addition to the media of protein kinase A or protein kinase C activators, or by inhibition of serine/threonine phosphatases. Although the data may currently be conflicting, it is likely that phosphorylation will turn out to be an important regulator of protein tyrosine phosphatase activity.

Another proposed mechanism of alteration in CD45 activity may include lateral movement of this receptor, causing changes in three-dimensional association with other molecules or/and putative substrates.

Whereas little is known regarding the function and regulation of the transmembrane protein tyrosine phosphatases, CD45 is unique in that its physiological role has been defined. It was established that CD45 is critical for the activation of T cells and may either directly or indirectly regulate protein tyrosine kinases which function at a very proximal point in the T cell receptor signaling cascade.

Analysis of tyrosine phosphoprotein patterns in CD45-deficient cells revealed an increase in the tyrosine phosphorylation of T cell receptor-associated protein tyrosine kinases Lck and Fyn. Phosphopeptide mapping studies demonstrated that both protein kinases were hyperphosphorylated at C-terminal tyrosine residues, sites of negative regulation for members of the Src-family protein tyrosine kinases.[183,248,288,346] The addition of CD45 to either Lck or Fyn in vitro resulted in dephosphorylation at the C-terminal tyrosine residue and a concomitant increase in their kinase activity.[274,275] These data demonstrated that Fyn and Lck appear to be among the specific substrates for CD45.

RECEPTOR GUANYLYL CYCLASES

Guanylyl cyclase receptors can be found within both the cytoplasm and membranes of most cells, where various ligands activate and elevate the level of intracellular messenger, cyclic GMP.[136] On this basis two classes of guanylyl cyclase receptors exist, those which contain no apparent transmembrane domain (soluble or cytoplasmic form), and those that contain a single putative transmembrane domain (particulate or membrane form).[76]

Four unique cDNA clones for membrane forms of guanylyl cyclase have been isolated from various mammalian cDNA libraries and have been designated as guanylyl cyclase (GC)-A, GC-B, GC-C, retinal GC and atrial natriuretic peptide (ANP)-clearance receptor. Both atrial natriuretic peptide and brain natriuretic peptide, C-type natriuretic peptide, and heat-stable enterotoxins/guanylin appear to represent the natural ligands for GC-A, GC-B, and GC-C, respectively.[135,211] The retinal GC has the same structural organization as other particulate forms but has been suggested to function as the photoreceptor guanylyl cyclase, not responding to soluble extracellular ligands.[345] The ANP-clearance receptor binds ANP and other natriuretic peptides but contains neither protein kinase-like nor cyclase catalytic domains. Some authors have suggested that it acts as a clearance receptor, whereas others have argued that it is a signaling molecule.[126]

The particulate forms appear to exist as homodimers or other higher-ordered structures and were shown to be composed of an extracellular peptide binding domain, a single transmembrane domain and intracellular protein kinase-like and catalytic domains (Fig. 3). An individual catalytic domain may not be sufficient for production of sufficient amounts of cyclic GMP by GC, and it has been confirmed that a dimer is a minimal catalytic unit. That is: (1) expression of individual subunits, each of which contained a consensus cyclase catalytic domain, resulted in no detectable soluble GC activity, whereas coexpression of the subunits in various cultured cell lines led to high cyclase activity;[57] (2) deletion mutants of the GC-A receptor, in which only the consensus cyclase catalytic domain was expressed, possessed enzymatic activity, but all the activity migrated coincident with homodimers;[376] (3) certain point mutations within the consensus catalytic domain of GC acted as dominant negative mutations by virtue of their ability to form inactive homomers or heteromers with wild-type subunits.[137]

Taken together, the data provide support for an argument that dimerization is required for expression of high cyclase catalytic

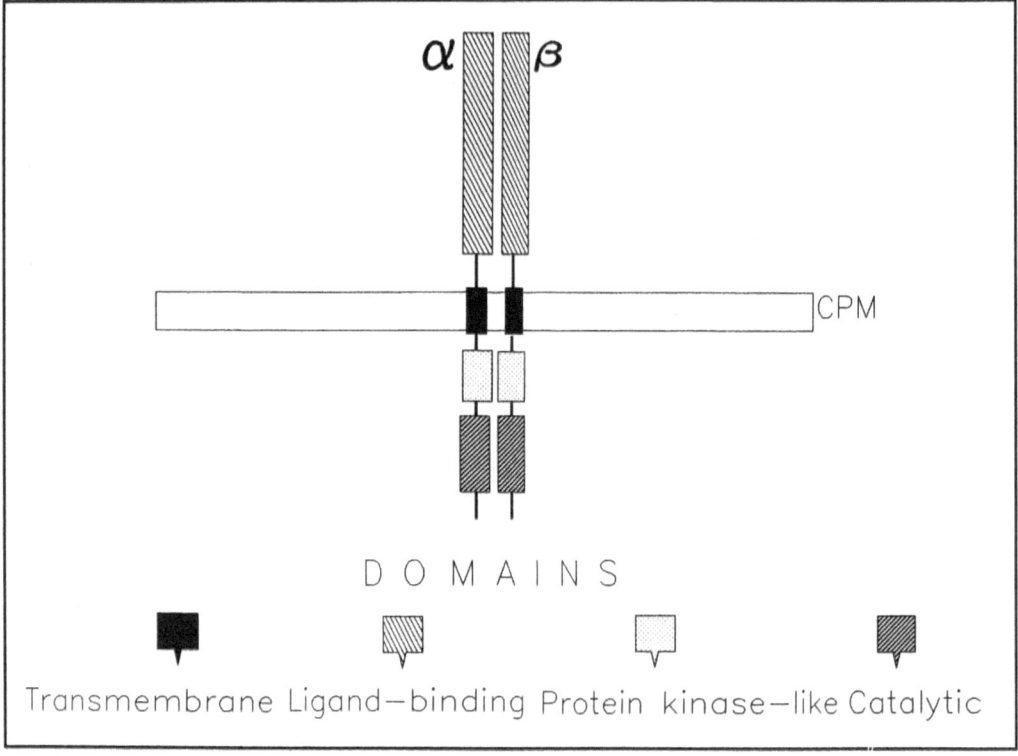

Fig. 3. Structural organization of receptor guanylyl cyclases. CPM, cytoplasmic membrane. Letters "α" and "β" stand for α and β subunits of the receptor correspondingly.

activity. However it is still not clear whether a single combined catalytic site is formed during the dimerization process or whether dimerization may be required for shared GTP binding sites or the appropriate folding of each subunit. It should be noted that receptor oligomerization was not linked with receptor activation, and the receptor was shown to exist as an oligomer in the absence or presence alike of its ligand.[177,235]

Several regulatory mechanisms have been described which modulate the activity of GC receptors. Among them is influence of ATP, which was demonstrated to strongly modulate the signal transducing activity of guanylyl cyclase receptors.[135,409] The site(s) of ATP binding is not certain but is most likely to be the protein kinase-like domain. In GC-A, deletion of the protein kinase-like domain resulted in a constitutively active cyclase that was no longer regulated by ANP or ATP. Likewise mutations within the protein kinase-like domain destroyed ATP regulation of the enzyme. Therefore it has been suggested that the protein kinase homology domain functions as a negative regulatory element and that ATP binding stabilizes a ligand-sensitive conformation, thus positively regulating the activity of the receptor.[75]

Another regulatory mechanism of guanylyl cyclase function might be phosphorylation of the enzyme, since phosphorylation of the GC-A receptor was observed on its serine and threonine residues. The significance of this finding is not known; however experimental data exist confirming that the desensitization of GC-A coincides with ANP-induced dephosphorylation of the receptor.[306] The fine mechanism by which dephosphorylation desensitizes the guanylyl cyclase receptor is however not known. One proposition exists that dephosphorylation changes conformational structure of the receptor and thus modulates affinity state of the enzyme.

RECEPTORS COUPLED TO PROTEIN KINASES

In addition to receptors that signal directly by phosphorylation/dephosphorylation reactions, many receptors are devoid of enzymatic activity and are able to stimulate protein tyrosine phosphorylation indirectly, acting on non-covalently bound cytoplasmic protein kinase. An example of this kind of receptor is the TCR/CD3/CD4,CD8 complex.

The T cell receptor (TCR) consists of α and β chains (Fig. 4). Both α and β chains have two extracellular Ig-like domains. The distal domains carry the variable portions and form the recognition site for the major histocompatibility complex (MHC)/peptide conglomerate; the proximal domains are invariant. The α and β chains are linked by a disulfide bridge and are anchored in the cell membrane by transmembrane regions with very short cytoplasmic tails. The TCR shows molecular diversity, similar to that of antibody molecules, which confers specific recognition properties to this receptor and to the T cells expressing it.[113] The five additional polypeptides (γ, δ, ε, ζ and η) represent the CD3 complex and are noncovalently associated with α/β TCR. They have larger cytoplasmic regions and are confirmed to play a role in transmembrane signaling.[67] Three proteins of the CD3 complex, γ, δ and ε, exhibit Ig-like extracellular domains, whereas the ζ-chain and η-chain have very short external regions.[365] The three-dimensional arrangement and the exact stoichiometry of the subunits within the TCR/CD3 complex (e.g., how many TCR α/β, γ/δ, δ/ε, ζ/ζ, ζ/η or η/η pairs exist per complex) are not known.

It is assumed that the CD3 dimers are essential both for efficient receptor expression and signal transduction, whereas the α/β dimer dictates ligand-binding specificity. The other structures necessary for normal operation of TCR/CD3 represent molecules called CD4 and CD8 (Fig. 5). Both CD4 and CD8 are members of the Ig superfamily.[220] CD4 is a monomeric transmembrane polypeptide with four extracellular Ig-like domains. CD8 is a disulfide-linked heterodimer of α and β chains. Each chain carries a single Ig-like domain which is linked by a hinge region to its transmembrane part.

CD4 and CD8 serve to discriminate helper versus cytotoxic response in T cells. Both cytotoxic and helper T cells use the same pool of TCRs, but they typically respond to antigen associated with different types of MHC molecules. Antigen associated with class I MHC molecules is sensed by CD8 molecules and generally elicits a cytotoxic response, whereas antigen associated with class II MHC evokes helper response after contact with CD4.

Accumulated data suggest that the TCR/CD3 complex employs two main transduction pathways. One of them starts with the

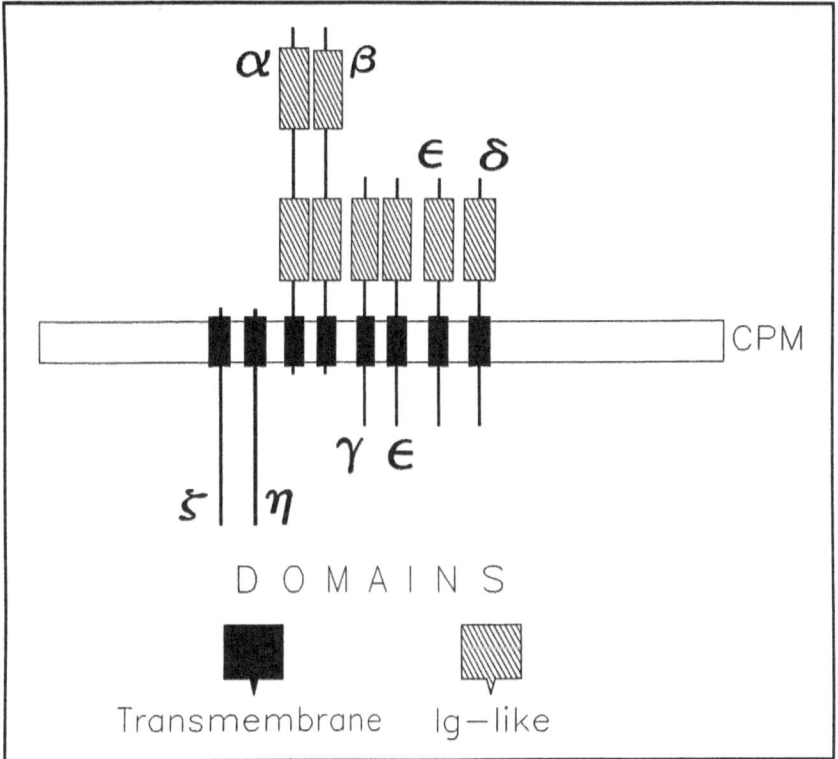

Fig. 4. Structural organization of the TCR/CD3 complex. CPM, cytoplasmic membrane. α-, β-, γ-, δ-, ε-, η- and ζ-chains are shown; however the exact stoichiometry of the subunits within the complex is not known precisely.

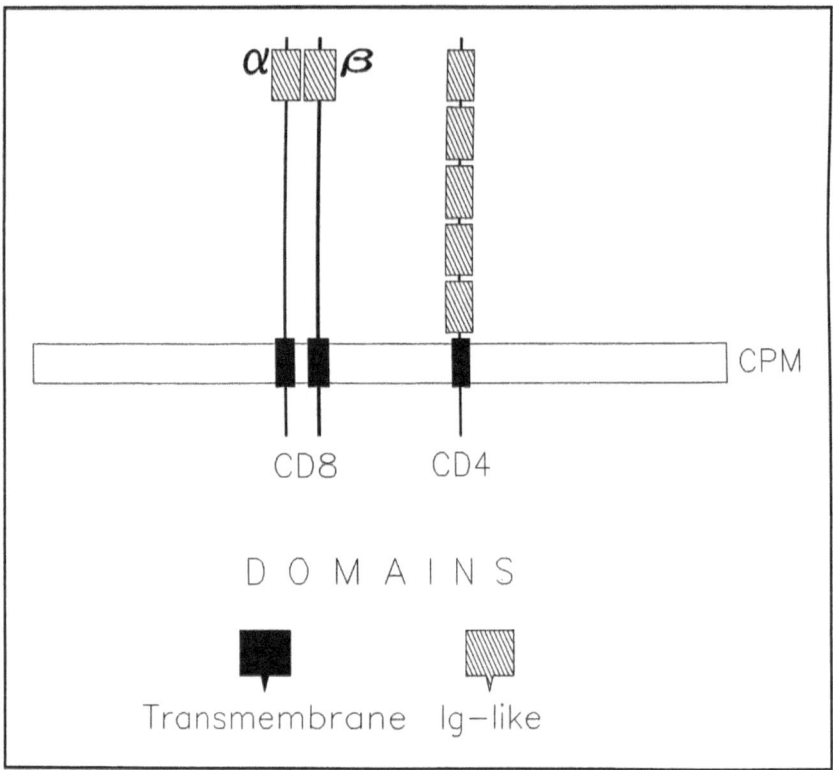

Fig. 5. Structural organization of CD8 and CD4 molecules. CPM, cytoplasmic membrane. CD4 consists of a single chain, whereas the CD8 molecule consists of α and β subunits.

cytoplasmic domain of ζ-family subunits. In addition, the cyto-plasmic tail of CD3-ε also has been shown to function as an au-tonomous signal transducer.[226] These data have led to the pro-posal that TCR/CD3 consists of at least two "transduction modules", the ζ and ε modules.

Activation of the multichain TCR/CD3 complex by antigen/MHC molecules induces stimulation of T cells. It is well documented that tyrosine phosphorylation of intracellular substrates is detected within seconds of TCR engagement, and tyrosine ki-nase inhibitors are able to block early signaling events in mature T cells. Based on these data, protein tyrosine kinases are consid-ered to be essential components of the proximal signaling cascade operated by the TCR/CD3 complex.[240]

Two classes of cytoplasmic protein kinases, members of the Src and Syk/ZAP-70 families, have been implicated in the func-tion of T cell antigen receptors. These classes of protein-tyrosine kinases display different structural composition which probably reflect their distinct functions. Members of the *src* gene family (Lck, Fyn and Yes) have an unique N-terminal domain with a myristoylated glycine residue at position 2 which is responsible for their membrane association. Obtained data indicate that p59^{fyn} is linked to the CD3-ζ chain, whereas the cytoplasmic tails of both CD4 and of the α chain of CD8 are associated with p56^{lck}.[327] ZAP-70, "ζ-chain associated 70 kDa protein," and Itk, the "IL-2 inducible T cell kinase," have features that distinguish them from Src-family tyrosine kinases. They lack amino-terminal myri-stoylation sites serving to anchor Src to the inner face of the mem-brane.[398] ZAP-70, which has two phosphotyrosine binding SH-2 domains, lacks a Src-characteristic SH-3 domain, whereas Itk con-tains one SH-2 and one SH-3 domain. The sequences of the ki-nase catalytic domains of ZAP-70 and Itk also differ from those of Src-family protein kinases.[68,347]

The process of initiating a signaling cascade from TCR/CD3 is in many details unknown. However recent data demonstrated a very complex network of phosphorylating reactions. One proposed scenario was that MHC-mediated crosslinking of the TCR with CD4 or CD8 initiates signaling by bringing p56^{lck} into proximity with the cytoplasmic regions of the TCR and CD3.[219] Antigen-induced TCR aggregation could favor phosphorylation of TCR tyrosine residues, allowing them to recruit other signal transducing proteins including cytoplasmic protein-tyrosine kinases ZAP-70 and Itk, through their SH2 domains. Such recruitment and activation of several protein tyrosine kinases as a result of triggering of the TCR-CD3 complex results in phosphorylation of a series of substrates. Among them are PLC-γ, 81 kDa ezrin, 100 kDa valosin-containing protein, as well as 67 kDa CD5 protein.[58,96,111,112] The bringing together of a panel of signaling molecules by activated T cell receptor allows initiation of different signaling cascades leading to complex physiological responses by T cells.

The regulation of tyrosine phosphorylation, accomplished by Lck and Fyn after stimulation of the TCR/CD3 complex, may be finely tuned by phosphorylation-dephosphorylation events on specific regulatory sites present on these protein kinases. At least two proteins have been shown to be engaged in this process, being allied at times with activated T cell receptor. These are a receptor-type protein phosphatase, CD45, and a 50 kDa protein kinase, Csk. Since the molecular target of CD45 is the negative regulatory site on p56^{lck} and p59^{fyn}, dephosphorylation of these tyrosine residues of the protein kinases leads to their activation. In addition, CD45 itself is phosphorylated upon T cell activation, which may affect its activity and association with the target Src protein kinases. This phosphorylation and attendant downregulation of CD45 is accomplished by Csk. Other targets for Csk activity are Lck and Fyn themselves.[27,286] Protein kinase p50^{csk} specifically phosphorylates p56^{lck} and p59^{fyn}, downregulating their activity. Thus p50^{csk} might determine a fine balance between p56^{lck}, p59^{fyn} and CD45 activities and thereby regulate the output signal via TCR.[365]

RECEPTORS COUPLED TO G PROTEINS

A large family of cell-surface receptors utilize guanine nucleotide-binding heterotrimeric G proteins for signal transduction. Hundreds of G protein-coupled receptors have been identified and the total number is assumed to exceed 1000.[152,287] Members of this receptor superfamily mediate cellular responses to diverse stimuli including light, odorants, glycoprotein hormones, lipids, alkaloids, peptides, nucleotides and biogenic amines. Several steps are involved in the process of G protein-coupled receptor activation: (1) creation of the signal by ligand binding; (2) transduction of the signal through the membrane; and (3) interaction with and activation of G protein.

Most of the known G protein receptors are composed of a single polypeptide chain forming seven putative transmembrane helices through the plasma membrane (Fig. 6). This topology generates an extracellular region (N-tail), at least three transmembrane loops and the carboxyterminal region (C-tail). In addition, palmitoylation of a cysteine residue for β2-adrenergic receptor [152] allows its membrane attachment and thus generates another intracellular domain. A cysteine residue is conserved at this position in many G protein-coupled receptors, suggesting a general functional role for this modification. Thus one class of lipidation of G protein-linked receptors—that is, palmitoylation—is necessary for their proper structural organization.

Palmitoylated proteins contain a 16-carbon saturated fatty acid group attached by a labile thioester bond to cysteine residues. Other fatty acyl chains may substitute for palmitoyl; therefore a more appropriate term would be S-acylated proteins. The acylation is posttranslational and its lability allows the process to be reversible, a unique property of this modification that gives cells the potential to control the modification state of the protein.

The extracellular N-terminal domain is of highly variable length and generally contains sites for N-linked glycosylation. For adrenergic receptors, the binding site for ligands has been localized to a cavity buried into the lipid bilayer and surrounded by the seven α-helical transmembrane segments. It is expected that binding of small ligands to other receptor types will probably be similar, but large ligands, due to steric conflict, appear to behave differently. That is, binding of massive glycoprotein hormones to a specific receptor has been shown to be accomplished via the extracellular N-tail in addition to extracellular loops.[294]

Special studies have shown the crucial importance of several modules for signal transduction from ligands to G proteins conferred by G protein-linked receptors. The N-R-Y (339-341) motif

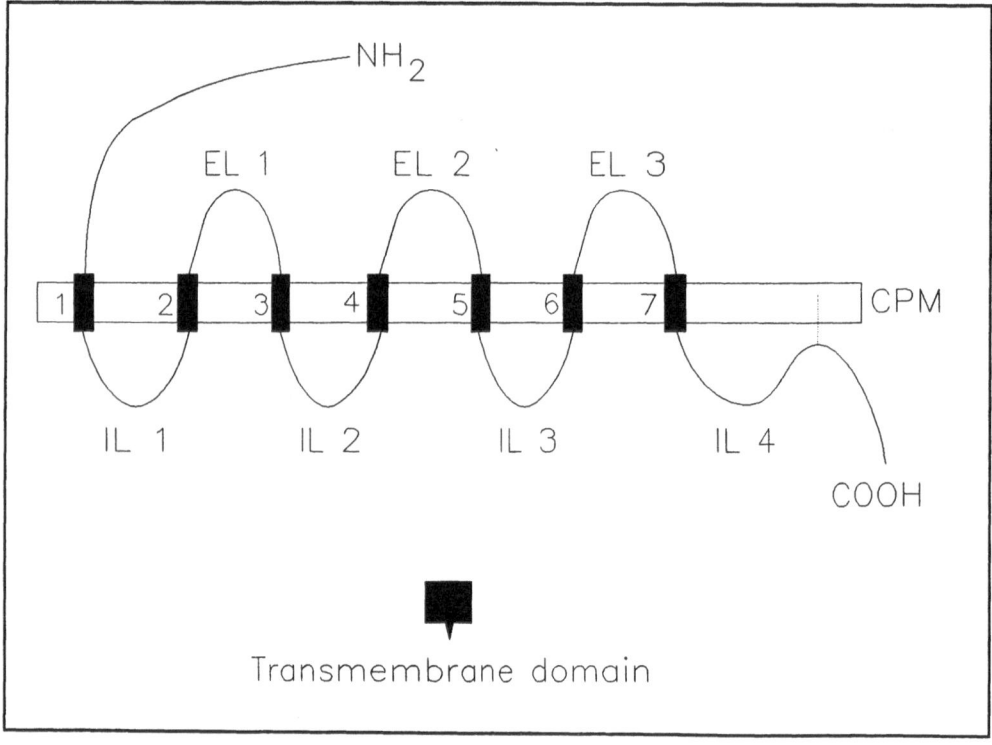

Fig. 6. Structural organization of G protein-coupled receptors. CPM, cytoplasmic membrane; EL 1-3, extracellular loops 1-3; IL 1-4, intracellular loops 1-4. Palmitoylation of a cysteine residue near the C-terminus allows its membrane attachment and generates intracellular loop 4.

at the cytosolic end of helix III and the loops between the helices III-IV and V-VI as well as the C-terminal domain have been shown to be important for G protein coupling and activation.[287] The other domain critical for signaling is represented by a "polar pocket", a space between the intracellular halves of helices I, III, VI and VII. However it should be noted that related domains can have dissimilar functions in different receptors.

Arginine residue-340, which is located near both the polar pocket and the cytosol, is hypothesized to participate in a "turning on-off" of the receptor. The receptor is "off" when the N340 side chain is in the polar pocket; and "on" when the N340 side chain is shifted toward the cytosol where it can reach the G protein. This switching is a result of a ligand binding and is caused by a subsequent rearrangement of the polar pocket. According to this scheme, agonist binding triggers a conformational alteration, such as the rotamer shift of one or more residues, that disturbs the polar pocket, pushes out the N340 side chain and favors G protein contact. This mechanism can also work in the opposite direction: G protein can contact the N340 side chain, which triggers a reversed chain of events leading to modified affinity for the ligand.

The activity of G protein-coupled receptor could be regulated by covalent modification reactions in addition to ligand binding. Thus the intracellular loop III and the C-tail generally bear potential sites for phosphorylation by protein kinase C or cAMP-dependent protein kinase. These phosphorylation sites could be important for the regulation of the receptor functions, although their functional significance has only been proven in a limited number of instances, such as the β-adrenergic receptors. The intracellular C-terminal segment in most cases is rich in potential sites for phosphorylation by β-adrenergic receptor kinase (βARK) or βARK-related serine/threonine kinases. These protein kinases phosphorylate the activated conformation of the receptor, which leads to the binding of a second cytosolic protein, termed arrestin in research on the retinal system and β-arrestin in the β-adrenergic receptors. These proteins bind to the phosphorylated forms of the receptors and then sterically restrict their interaction with G proteins, thus leading to functional deactivation of the system.[222]

RECEPTORS COUPLED TO ION CHANNELS

Among receptors which regulate distinct ion channels of a cell, the most thoroughly studied include 1,4,5-inositol triphosphate receptors (IP$_3$Rs).[30,262] The function of this kind of receptors includes the regulated release of Ca^{2+} ions in response to 1,4,5-inositol triphosphate (IP$_3$), a potent intracellular signaling molecule.

The physiological significance of Ca^{2+} is enormous. Calcium ions act on a wide variety of targets, e.g., phospholipases, guanylyl cyclases, nitric oxide synthases, Ca^{2+}-binding proteins such as calmodulin and calbindin, protein kinases, protein phosphatases, proteinases, ion channels such as Ca^{2+}-dependent K$^+$ and Cl$^-$ channels, receptors such as IP$_3$R itself and another intracellular Ca^{2+}-releasing channel—ryanodine receptor—cytoskeletal proteins such as actin, transcription factors etc.[127]

IP$_3$Rs were shown to be highly enriched in cerebellar Purkinje cells but are also expressed in the hippocampus, striatum and cerebral cortex where IP$_3$ appears to work.[128] Inositol triphosphate receptors were not only highly expressed in the brain, but also expressed in the smooth muscles of arteries, oviduct and uterus, in the contractile smooth muscle of the aorta, intestine and esophagus. Electron microscopic observation demonstrated that the receptors are localized mostly on the smooth endoplasmic reticulum, but are also present in the rough endoplasmic reticulum and outer nuclear membrane.[261] Ca^{2+} signals mediated by these nuclear IP$_3$Rs may be involved in nuclear functions, including the transcriptional activation of immediate-early genes. In Purkinje cells stacked endoplasmic reticulum cisternae, highly enriched in IP$_3$Rs, are closely opposed to the mitochondria. It should be stressed that of the known second messengers only the Ca^{2+} signal appears to originate from such cisternae and could be transferred into the mitochondrial matrix and modulate certain key enzymes in ATP synthesis.[100,317] In nonneuronal cells several lines of evidence indicate the presence of the plasma membrane IP$_3$Rs, which probably function as an IP$_3$-activated Ca^{2+}-permeable channel.[262]

The inositol 1,4,5-triphosphate receptor was first characterized in 1979 as a protein called P400 that was abundant in the cerebellum but was virtually absent in the cerebella from Purkinje cell-deficient mutant mice,[263] and purified in 1988 independently by two groups.[238,364] Three distinct receptor types (types 1, 2, and 3) in the IP₃R family have since been molecularly cloned. Each receptor type was characterized by differences in the IP_3-binding and Ca^{2+}-releasing activity, suggesting the existence of differential IP_3/Ca^{2+} signaling. Within this family, the IP₃R 1 is the most thoroughly characterized molecule.

The IP₃R 1 is structurally divided into four parts: the large N-terminal cytoplasmic arm, the regulatory domain, the membrane spanning and pore-forming regions and the short C-terminal cytoplasmic tail (Fig. 7). The N-terminal arm (about 650 amino acid residues) is highly conserved among different species. Deletion of any small fragment within the region abolished IP_3 binding activity, suggesting that this region is important for IP_3 binding.[260] The N-terminal arm is enriched with positively charged arginine and lysine residues and it binds heparin.

The regulatory portion of the receptor functions as the transducing module involved in the connection of IP_3 binding to channel opening and contains binding sites for various modulators such as Ca^{2+}, calmodulin and ATP. This part of IP₃R is subject to modifications by various protein kinases. There are two sites phosphorylated by protein kinase A, one site phosphorylated by protein kinase G and potential sites for phosphorylation by Ca^{2+}/calmodulin-dependent protein kinase 2 and protein kinase C.[127] Thus, different second messenger transducing cascades, i.e., cAMP (via protein kinase A), diacyl glycerol (via protein kinase C) and cGMP (via protein kinase G) converge on this domain. In addition, IP_3-induced Ca^{2+} releasing activity is regulated by cytosolic Ca^{2+} concentration in a biphasic dose-dependent manner. Thus subtle regulation of the transducing action by "cross-talk" among these diverse signaling pathways appears to cause differential IP_3 / Ca^{2+} signaling.

The multiple membrane-spanning domains and one putative pore-forming sequence are thought to constitute the channel domain.[258] The C-terminal region of the IP₃R is considered to play a role in Ca^{2+} release, since a monoclonal antibody that recognizes the C-terminus of the type-1 IP₃R was found to suppress the Ca^{2+} channel activity.[277]

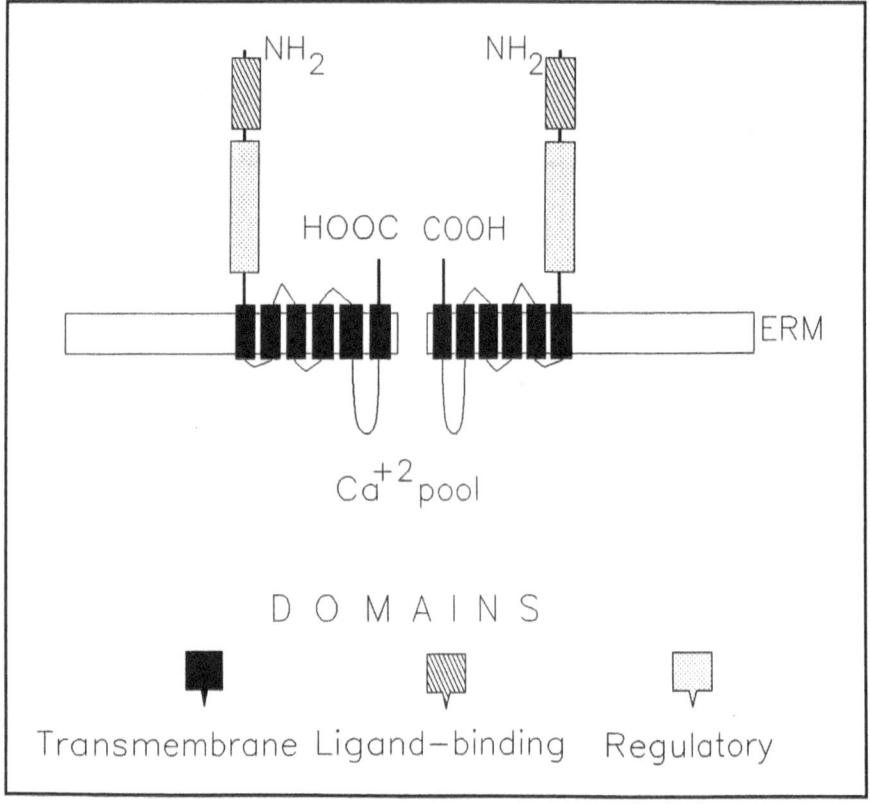

Fig. 7. Structural organization of 1,4,5-inositol triphosphate receptor. ERM, endoplasmic reticulum membrane. The receptor is thought to consist of four identical subunits. Only two subunits are shown for the sake of simplicity.

Cross linking experiments and electron microscopic observations have demonstrated that the receptor forms a homotetrameric structure. Membrane spanning domains and the C-tail are apparently required for this intermolecular association. Each subunit of IP_3R binds one IP_3 molecule with no cooperativity in inositol triphosphate binding as revealed by the IP_3 binding kinetics.

GPI-ANCHORED PROTEINS

Some cell-surface proteins are anchored in membranes by covalent attachment to glycosylphosphatidylinositol (GPI) rather than by conventional transmembrane spans.[53] In such a case, the extracellular protein is coupled via an amide link to ethanolamine, which forms a phosphodiester linkage to mannose (Fig. 8). This sugar is part of a glycan containing a glucosamine residue linked to the inositol ring of phosphatidylinositol (PI). A number of GPI-anchored membrane proteins may be involved in signaling events and certain GPI lipids have been postulated as precursors in the transmembrane signaling pathways.[51]

An understanding of the importance of GPI-anchored proteins has arisen mainly from the studies on these molecules in T cells which express GPI-anchored proteins abundantly. It was shown that the mechanisms of T cell activation through the T cell receptor and through GPI-anchored proteins were very similar. Consequently, antibody-mediated clustering of TCR or GPI-anchored proteins stimulated downstream signaling events which were closely related in the two pathways.[53,320] In both cases binding of a ligand (or antibody) to the cell-surface protein was followed by activation of intracellular kinases such as $p56^{lck}$, $p59^{fyn}$ or $p62^{yes}$.[176] Stimulation of tyrosine kinase activity and cell activation through TCR required the leukocyte-specific protein tyrosine phosphatase CD45. Likewise, a close association between CD45 and the GPI-anchored protein Thy-1 has been demonstrated by chemical crosslinking and it seems likely that the phosphatase may activate kinases that are allied with GPI-anchored proteins.[357]

It remains undetermined how molecules associated with opposite leaflets of a membrane bilayer communicate with each other. One proposition is that a transmembrane "linker" protein medi-

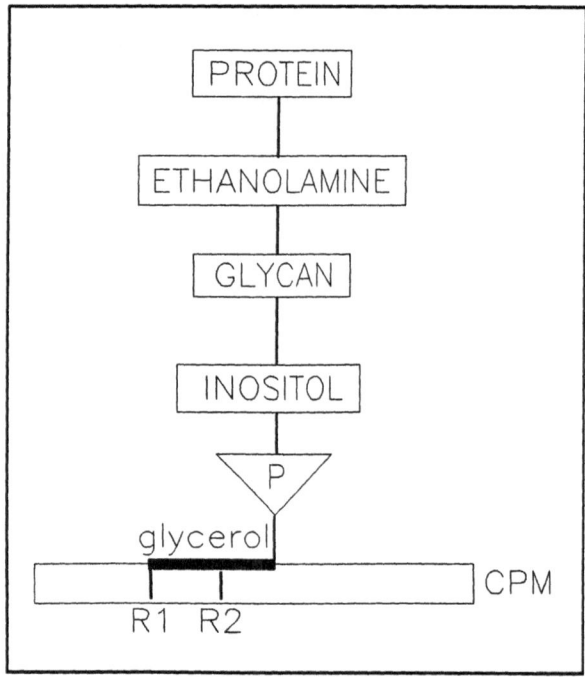

Fig. 8. Structural organization of glycosylphosphatidylinositol-anchored proteins. CPM, cytoplasmic membrane; R1, R2, fatty acid residues; P, phosphate group.

ates the interaction between the GPI-anchored protein and protein kinase. In fact, there may be a family of these linker proteins,
and different members of the family may associate with the different Src-like kinases in a specific manner. This might help explain
the fact that in the three studies[343,357,375] either p56[lck], p59[fyn] or
both proteins were found to be associated with GPI-anchored
proteins.

Another possibility exists that GPI-anchored proteins may not
bind directly to transmembrane linker proteins in order to con-
nect with intracellular kinases. Rather, GPI-anchored proteins seem
to have an affinity for clusters of specialized microdomains of associated lipids in the outer leaflet of the bilayer.[53,348] Some transmembrane proteins may also associate with these lipid domains.
Binding of the kinases to the cytoplasmic domain of such proteins
could lead to a close association with GPI-anchored proteins in
the opposite leaflet of the bilayer. It is possible also that kinases
do not bind to transmembrane proteins, but associate directly with
lipid microdomains.

Bamezai and colleagues[16] showed that immobilization of antibodies directed against GPI-anchored proteins (but not of TCR-
specific antibodies) prevented activation, and that GPI-anchored
proteins were internalized from the surface of T cells. These results suggested that the proteins must be internalized for activation to occur. The authors found also that internalization of
GPI-anchored Ly-6A.2 and Thy-1 proceeded by a mechanism distinct from receptor-mediated endocytosis in clathrin-coated pits.
In subsequent studies another group of researchers has shown that
GPI-anchored proteins could be transiently internalized via non-
clathrin-coated membrane invaginations called caveolae by a process termed potocytosis.[9] Investigating the uptake of folate after
binding to the GPI-anchored folate receptor as a model system,
these authors have proposed that potocytosis is a general mechanism for transport of small molecules into the cytoplasm. Thy-1
and Ly-6A.2 may be internalized by this mode. The fate of the
proteins once in caveolae and the role of internalization in signaling are still however not clear. Perhaps GPI anchors are cleaved,
leading to the generation of signaling molecules that are trans-
ported into the cytosol.

IMMUNOPHILINS

Immunophilins were first described as cytosolic receptors for certain immunosuppressive drugs. Among the latter was cyclosporin A, an inhibitor of T cell activation and one of the widely used therapeutic agents for prevention of graft rejections in clinical transplantations. Another ligand for immunophilins is macrolide FK506, whose mechanism of inhibition of T cell activation is likely to be similar to that of cyclosporin A but the degree of inhibition is 10-100 times higher, and rapamycin, whose mechanism of inhibition is supposed to be different from that of cyclosporin A and FK506.[337] Immunophilins which bind the corresponding immunosuppressants were termed cyclophilin, FK506-binding protein (FKBP) and rapamycin-binding protein accordingly.

Studying enzymatic properties of cyclophilin and FK506-binding protein, it was disclosed that these proteins are representatives of peptidyl-prolyl *cis/trans*-isomerases, enzymes which catalyze the interconversion of the *cis*- and *trans*-rotamers of the peptidyl-prolyl amide bond of protein substrates.[259,337] During special experiments it was shown that immunosuppressive compounds bound immunophilins and inhibited peptidyl-prolyl *cis/trans*-isomerase activity. It was also proposed that this event is probably linked to mechanisms of immunosuppression.

Cyclophilin is an abundant protein with a molecular mass of around 18 kDa and pI = 9.3. Human cyclophilin consists of eight β-sheets interrupted by α-helical regions. Experimental data suggest that the β-sheets form a groove which includes the amino acids important in peptidyl-prolyl *cis/trans*-isomerase activity and cyclosporin A binding. FK506-binding protein is characterized by molecular mass of ca. 12 kDa and pI = 8.9. The human FKBPs contain a central domain of five β-sheets interrupted by α-helix. Besides FK506, FKBP is able to bind rapamycin with very high affinity.[331]

In addition to ligand binding, immunophilins are presumably subjected to regulation by covalent modification. For instance, polypeptide chains of both cyclophilin and FKBP contain consensus sequences for phosphorylation, prenylation, myristoylation and N-glycosylation.[154] As a result of these reactions, fine tuning of immunophilins' biological activity can be accomplished in a cell.

Using T cells as a model it was shown that one result of cyclosporin A and FK506 binding to the corresponding immunophilin member was inhibition of signaling from different receptors, including T cell receptor, AP-3 and Oct-1. It resulted in a cascade of events, among them decreased mRNA synthesis for certain genes including *c-myc, hsc-70, hsp-70, krox20*.[154] In addition, both compounds, but not rapamycin, were shown to be potent inhibitors of apoptosis of T cell hybridomas induced by antibody to T cell receptor-CD3 complex. Rapamycin has been demonstrated to inhibit lymphokine receptor-associated signaling pathways.[337]

Several lines of evidence suggested however that inhibition of only rotamase activity of immunophilins by immunosuppressive compounds was not sufficient to mediate the effect. The distinct mechanisms of action of immunophilins and immunosuppressants are however not known in detail to date. It was first proposed that immunophilin-drug complexes assist protein folding in vivo, thus acting as foldases. In such a case immunophilin receptor bound to FK506, cyclosporin or rapamycin may communicate with certain signaling molecules in its close vicinity and change their stereo conformations. The proposed outcome is altered signaling events in eukaryotic cells, which are controlled by such signaling molecules. Second, immunophilins may function analogously to G proteins, in the sense that intrinsic peptidyl-prolyl *cis/trans* isomerase activity may perform analogously to GTPase activity and regulate switching on-off of a transduced signaling event. Third, immunophilins coupled to a distinct signaling pathway may function as presenting molecules for immunosuppressants by analogy to major histocompatibility complex molecules which are able to present numerous antigens to receptors on immune cells.[337]

Structural analysis of immunophilins' ligands supplied some illustrations of the latter hypothesis. During these investigations data have been obtained indicating the possibility of at least two domains in FK506 and rapamycin molecules. The domain that binds immunophilin has overall structural homology with the prolyl-containing site of a potential target protein for this rotamase and represents the pseudosubstrate for peptidyl-prolyl *cis/trans* isomerase. The other domain, called the "effector domain," was

able after presentation to the effector molecule by an immunosup-
pressant-immunophilin complex to interfere with signaling events
in eukaryotic cell.

In addition to the cyclophilin and FKBP described above, other
members of the immunophilin family have been found. These are
45 kDa cyclosporin A-binding phosphoprotein, 60 kDa and 80 kDa
phosphoproteins, 30 kD'a and 13 kDa FK506- and rapamycin-binding proteins and some others. It is certain that the number of
members of the immunophilin family will grow and help the precise determination of biological significance of immunophilins and
their ligands.

CHAPTER 2

HETEROTRIMERIC
G PROTEINS

A considerable number of proteins incorporate GTP in a regulatory cycle of their activity. These proteins are able to bind GTP, possess intrinsic GTPase activity and undergo GTP-dependent conformational changes critical for their functioning. This superfamily of GTP-binding proteins consists of several families including translational factors, heterotrimeric GTP-binding proteins (G proteins), protooncogenic Ras and Ras-related proteins, tubulins and some others.[197] G proteins stand apart from other members of the superfamily of GTP-binding proteins due to their heterotrimeric structure and their function in coupling of specific cell surface receptors to effector molecules. Initially they were discovered as critical intermediates in visual transduction in photoreceptor cells of the retina and in hormonal stimulation of cAMP production in the liver. However in subsequent studies G proteins became recognized as signal transducers in a vast number of signaling pathways.[35,354]

G proteins are composed of three distinct gene products: α, β and γ subunits. These subunits are able to form a heterotrimer associated with the specific receptor. Following receptor activation by ligand binding, the α subunit dissociates from this heterotrimeric complex leaving the β and γ subunits noncovalently associated in the βγ subunit.[353] It is primarily the α subunits which are believed to confer specificity in receptor-G protein coupling and to determine the type(s) of signal transduction pathways triggered by agonist binding.

Effector structures regulated by different subtypes of G proteins include adenylyl cyclases, Ca^{2+}, Cl^-, Na^+ and K^+ channels, cGMP phosphodiesterases, phosphatidylinositol-specific phospholipases Cβ (PI-PLCβs), phosphatidylinositol kinases, Ras-associated regulatory proteins and protein kinases. While the general way of activating a specific effector is via an α subunit, several lines of evidence exist that the $\beta\gamma$ complex is also involved in activation of these enzymes in certain cases.[284] Effectors regulated by $\beta\gamma$ dimers may include adenylyl cyclases, phosphatidylinositol-specific phospholipase Cβ, K^+ channels and phosphatidylinositol kinases.[283] There is evidence that G$\beta\gamma$ subunits contribute, the down-regulation of the corresponding receptor, thus participating in a negative feedback mechanism. In addition, recent evidence suggests that activation of phospholipase A_2 may be accomplished by a single β subunit as well.[298]

SUBUNIT COMPOSITION OF HETEROTRIMERIC G PROTEINS

To date, more than 20 distinct α subunits are known that can be divided into four subfamilies based on the degree of their sequence homology. These are α_s, α_i, α_q and α_{12} subfamilies.[284] It is agreed that the name of a given G protein depends on its α subunit contents. Among the different G protein α subunits known to exist, G_s and olfactory-specific G_{olf} α subunits stimulate adenylyl cyclases. G proteins of the α_i class regulate activity of phosphodiesterases, Ca^{2+} channels, inhibit adenylyl cyclases and may cause weak activation of PI-PLCs, depending on the cell type. G proteins containing α_q subunits strongly promote activity of β-type phosphatidyl inositol-specific phospholipases C. The effectors of G_{12} α subunits remain unstudied.[161,283]

The α subunits range in size from 350 to 395 amino acid residues. Biochemical, immunochemical and molecular biologic approaches have helped define functional domains of Gα. An amino-terminal 1-2 kDa region is directly involved in binding to the $\beta\gamma$ dimer. The carboxy-terminus is critical for receptor interaction. Recent evidence, based on studies of a chimerical α subunit, suggests that the effector interaction domain is also within the carboxyterminal region, perhaps proximal to the receptor interaction domain. Another part which is of critical importance for the functioning of α subunits is the GTPase domain. No definite function has been assigned to a large domain proximal to the effector domain which shows substantial sequence diversity among various α subunits. One suggestion is that this domain regulates activity of the GTPase domain of the α subunit and functions as an endogenous GTPase-activating protein (GAP) comparable to the GAPs known to regulate Ras proteins.

Together with the α subunit, the βγ heterodimer serves an essential role in G protein-receptor coupling. Five β subunits and seven γ subunits have been described so far.[284] Whereas β subunits appear to be closely related according to their amino acid sequences, γ subunits are considerably more diverse and functional differences of βγ complexes may be primarily due to the γ subunit.

The β subunits are about 430 amino acid residues in length.[36] They show a repetitive segmental structure made up of adjacent homologous domains, each about 43 amino acid residues long, and are delineated by a tryptophan-aspartic acid motif. The functional significance of this repetitive structure is not yet apparent. The γ subunits are about 70 amino acid residues in length. The domains involved in association of β and γ subunits and in association of the βγ heterodimer with α subunits have not been elucidated in full detail. Putative receptor and effector interactions domains are also undefined. There is no rigorous evidence, in fact, that the βγ heterodimer interacts directly with either receptors or effectors.

Covalent modifications by lipids are critical for membrane association of α and γ subunits and their stability. Several α subunits are known to undergo cotranslational myristoylation as follows: cleavage of the amino terminal methionine residue and amidation of the next residue (glycine) by myristate (14:0). There is no absolute consensus sequence for myristoylation, but a glycine residue at position 2, immediately following methionine, is absolutely required, and a hydroxyamino acid (typically serine) at position 6 is preferred.[355] A specific enzyme—N-myristoyl-transferase—participates in this process. In addition, all α subunits have been shown to undergo palmitoylation near the N-terminus.

These lipid modifications influence interactions between the proteins in the heterotrimer or between α subunit and the plasma membrane.[63,407]

Since not all lipid-modified proteins are localized to the plasma membrane, additional factors must be involved in membrane attachment. It is tempting to speculate that the βγ complex serves as such a receptor for α subunits. However, the data showing the independent membrane association of α subunits argues against an obligatory role for the βγ complex as an α subunit anchor.[355]

Certain types of Gα are subjected to phosphorylation or have consensus sequences for phosphorylation by protein kinase C, protein kinase A and protein kinase G.[283] However the functional importance of this kind of modification is unclear.

The γ subunits of G proteins usually undergo three sequential modifications characteristic of proteins possessing a carboxyl-terminal CAAX motif (C, cysteine; A, aliphatic; X, any amino acid residue): isoprenylation of the cysteine fourth from the carboxyl terminus (farnesyl, 15-carbon chain, when X is serine; geranylgeranyl, 20-carbon chain, when X is leucine); proteolysis of AAX residues; and methylation of the resultant carboxyl group. Prenylation and proteolysis of AAX residues appear to be irreversible modifications performed in the cytoplasm. In contrast, methylation occurs in a membrane-bound compartment and may be reversible as well as regulated.

Assembly of the βγ complex was shown to be not dependent on prenylation of the γ subunit. In contrast association of a βγ with an α subunit to form the trimer is dependent on γ prenylation. Although again the exact mechanism through which the lipid contributes to α-βγ interactions is unclear, it is possible that the interaction between the lipids attached to α and βγ is one of the determinants in trimer formation.[64]

GTPASE CYCLE OF HETEROTRIMERIC G PROTEINS

The basis for accurate functioning of a G protein is its interaction with GTP-GDP molecules, known as the GTPase cycle (Fig. 9). The inactive G protein exists as a coupled to receptor heterotrimer with GDP tightly bound to the α subunit (Fig. 9, upper left panel). The resting G protein-associated receptor shows

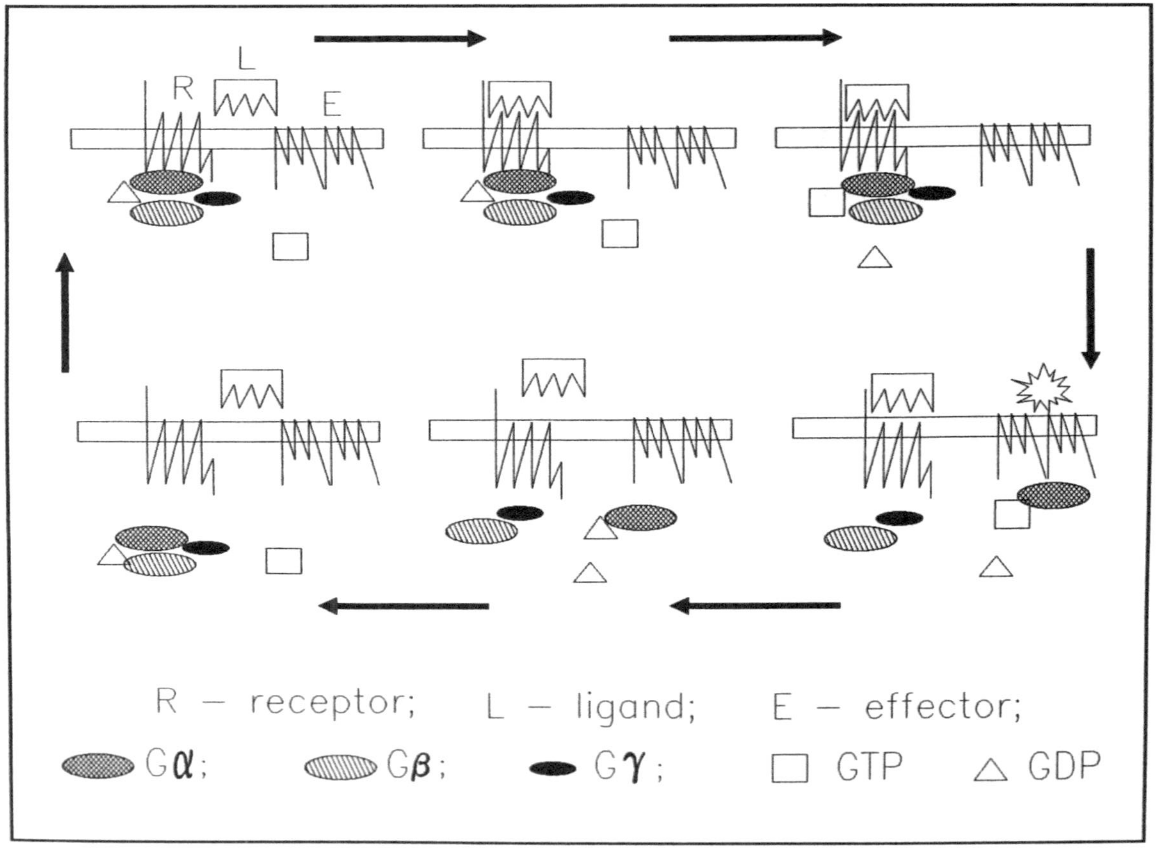

Fig. 9. The GTPase cycle of G proteins. The resting condition of the receptor-G-protein complex is shown on the left upper panel, whereas an episode of target enzyme activation is shown on the right lower panel. For details see text.

high affinity for agonist and enters an activated state following ligand coupling (Fig. 9, upper middle panel). The function of activated receptors is to catalyze exchange of GTP for GDP on G protein (Fig. 9, upper right panel). GTP binding activates the α subunit, reduces its affinity for the βγ heterodimer, leads to dissociation of the heterotrimer and lowers receptor affinity for agonist. The free activated α subunit, and in some cases free βγ complex or single β subunit, now interacts with and regulates effector molecules (Fig. 9, lower right panel). Generally, α subunit activation is transient and is terminated by the intrinsic GTPase activity of the subunit (Fig. 9, lower middle panel). Upon hydrolysis of GTP to GDP, the α subunit is deactivated and regains high affinity for the βγ heterodimer, leading to reassociation of the heterotrimer (Fig. 9, lower left panel). The latter is the form capable of coupling to the receptor and thereby reentering another round of the GTPase cycle.

The absolute requirement of G proteins for GTP, and the opposing effects of GDP and GTP on the activity of GTP-binding proteins, suggest that rephosphorylation of GDP arising from GTP hydrolysis is essential for their function. It is known that extramitochondrial GTP regeneration can be accomplished through several transphosphorylation reactions catalyzed by (1) guanosine triphosphate monophosphate kinase (2GDP = GMP + GTP), (2) guanylyl kinase (GDP + ADP = GTP + AMP), and (3) nucleoside diphosphate kinase (GDP + NTP = GTP + NDP). The latter enzyme is ubiquitous and in most tissues its activity is 10- to 100-fold greater than the activity of the nucleotide monophosphate kinases.[293]

Nucleoside diphosphate kinase is found in preparations enriched in GTP-binding proteins, such as microtubule proteins, rod outer segments, ribosomes and plasma membranes.[289] Since nucleoside diphosphokinase activity in the membrane has been found to form a complex with G proteins, this enzyme could serve to maintain high levels of GTP in the immediate vicinity of the G protein or could even act on GDP bound to the α subunit without requiring actual exchange of GTP for GDP.

In addition to this well defined mechanism of G protein activation another parallel pathway has been proposed.[193] According to this hypothesis, the β subunit accepts the phosphate group generated through GTPase activity of Gα and transfers it to the neighboring α subunit. It leads to an activated state of the second αβγ trimer. It was also speculated that such phosphorylated Gβ represents an activated state of the molecule by analogy to activated GTP-bound Gα.

CHAPTER 3

LOW MOLECULAR WEIGHT GTP-BINDING PROTEINS

The *ras* gene for low molecular weight GTP-binding proteins was discovered as an oncogene in mammalian tumors more then 10 years ago and its central importance in many signaling pathways has since been established.[17] Presently, the Ras superfamily is composed of more than 50 mammalian proteins of 20-25 kDa in size which are grouped into several distinct families.[191]

While the precise function of Ras proteins in many cases remains to be elucidated, many details of the biochemical pathways that are regulated by Ras proteins and how Ras mediates these intracellular signaling cascades, have been brought to light.[199] Recent genetic and biochemical studies have provided a detailed scheme which portray the Ras proteins as critical intermediates in signaling along various pathways in eukaryotes. Analysis of the level of Ras-bound GDP and GTP in intact cells has shown that various extracellular stimuli could trigger the activation of Ras in various types of cell lines. Signal transduction pathways that have been found to involve Ras protein include growth inhibition or stimulation and transformation of cells; proliferation and differentiation of hematopoietic and neuronal cells; activation of immunocompetent cells; cytoskeletal rearrangements and intracellular organelle traffic and many others.[251,330]

STRUCTURE OF RAS PROTEINS

An advance in understanding the molecular mechanism of p21^{ras} function came from the analysis of functions and X-ray structure of mutated proteins.[399] During these investigations certain areas of Ras proteins were shown to be the most important for normal operation of G proteins.

Among the mutated residues most frequently resulting in tumor formation were G12 and Q61. These mutations caused severe reduction of GTPase activity of p21^{ras}. On the other hand oncogenic p21 was still able to bind GTPase activating protein, and, as with the wildtype Ras, binding is dependent on the presence of GTP. These amino acid residues are proposed to mediate interaction of p21 with phosphate groups of GTP and to participate in GTP hydrolysis by activating nucleophilic attack on the γ-phosphate.

The other amino acid residue which is important for proper functioning of Ras proteins is S17. This amino acid is believed to be involved in Mg^{2+} coordination and participates in contact with the Ras nucleotide dissociation stimulator, thus regulating the GDP dissociation rates. Amino acid residues 28-40 were found to be situated in a loop which appeared to be detached from the protein. Through mutagenesis experiments this region has been found to be necessary for GTPase activation by GTPase activating protein as well as for biological activity in general of Ras. This loop has been defined as important for interaction with a downstream effector and was therefore named the "effector region." The effector region was subjected to significant conformational perturbations following GTP hydrolysis and thus could participate in switching on-off of signal transduction.

Ras proteins are only active in cellular signal transduction when they are localized at the inner surface of the plasma membrane or on other membranous structures after attachment of fatty acids. Several kinds of posttranslational lipid modifications have been described. One of such modifications important for Ras protein activity is isoprenylation of the CAAX tail (C = cysteine, A = aliphatic, X = any amino acid, most often serine or methionine residues). In addition, several representatives of the Ras family are palmitoylated on one or more cysteine residues within the C-terminal hypervariable region for cooperation with the farnesyl moiety in membrane attachment.

GTPASE CYCLE OF RAS PROTEINS

Ras protein binds 1 mol of GDP or GTP per 1 mol of protein and hydrolyzes the bound GTP to GDP and inorganic phosphate. The significant structural similarity with G protein α subunits suggests that Ras functions as a signal transducer in a manner similar to other signal-transducing GTP-binding proteins.

Indeed low molecular weight GTP-binding proteins undergo two alternative conformational transitions depending on the phosphate potential of bound nucleotide. The GTP-bound conformation is "active" in the sense that it can interact with a target molecule which transduces signal downstream. On the other hand, the GDP-bound conformation is referred to as "inactive" since it cannot stimulate the downstream target. However it can interact specifically with an upstream regulator. The two conformations can be converted into each other in two ways. One is the nucleotide exchange reaction, in which GDP dissociates from protein-GDP complex and a ligand-free protein immediately binds GTP, which is more abundant than GDP. The other reaction is the hydrolysis of protein-bound GTP to GDP and inorganic phosphate due to the intrinsic GTPase activity of the GTP-binding protein. During this process the GTP-bound active conformation is converted to the GDP-bound inactive form.

The interconversions of Ras protein-GDP to Ras protein-GTP and vice versa are finely regulated by special proteins, the proper functioning of which is absolutely necessary for normal signaling through Ras. Three distinct classes of such regulatory proteins have been identified and molecularly cloned. GTPase activating proteins (GAPs) and guanine nucleotide dissociation inhibitors (GDIs) negatively regulate Ras proteins by stimulating the hydrolysis of GTP and inhibiting the GDP-GTP exchange respectively. In contrast guanine nucleotide dissociation stimulators (GDSs) positively regulate Ras proteins by triggering the formation of their active GTP bound form.[44,104,156]

GTPASE ACTIVATING PROTEINS

Several distinct and ubiquitously expressed mammalian GTPase activating proteins for Ras have been identified. These are p120GAP and neurofibromin (NF1GAP) which share both sequence homology and substrate specificity with each other.[395] In addition, an alternative splicing event generates a smaller version of p120GAP (p100GAP), that appears to be expressed only in the placenta and lacks the N-terminal hydrophobic sequences present in p120GAP. The function of this variant of p120GAP is not known. GTPase activating proteins are primarily cytosolic proteins capable of stimulating the GTPase activity of Ras proteins by more than 10,000 fold.[381] In addition to their regulatory activity on the GTP hydrolysis rate of Ras proteins, GAPs represent effector molecules which are capable of transmitting signals downstream.[118,199]

The molecular organization of p120GAP is studied in considerable detail. Structurally p120GAP is composed of two functional domains. The C-terminal domain, designated the catalytic domain, contains the Ras-binding domain and interacts with the Ras protein to stimulate hydrolysis of bound GTP.[3] The N-terminal domain is believed to regulate the activity of the catalytic domain.[45] Additional information from several laboratories demonstrated that the N-terminal region of the p120GAP could interact with putative downstream effectors of Ras function and thus could transmit some signals. For example, the N-terminal domain but not the full length protein, was capable of activating of transcription from a *fos* promoter. Similarly, p120GAP inhibited muscarinic K^+ channel activity in a Ras-independent manner.[242]

The structural GAP components of key importance are two SH2 and one SH3 domains, a calcium-dependent phospholipid-binding domain and pleckstrin homology domain.[243] These domains are necessary for signal-induced trafficking of GAP and are capable of modulation of its activity. Thus SH2 domains are involved in complexing with receptor and non-receptor tyrosine kinases following their stimulus-induced phosphorylation on tyrosine residues.[7,117] One of the present models suggests that the binding of Ras-GTP to the catalytic domain of p120GAP results in conformational change of the latter that exposes the N-terminal SH2/SH3 domains, thereby promoting their interaction with certain downstream target(s) to transmit the Ras signal.

GUANINE NUCLEOTIDE DISSOCIATION STIMULATORS

Ras proteins have a very low intrinsic exchange activity which keeps them in the GDP bound state under basal cellular conditions.[329] Consequently, it is speculated that regulatory factors which enhance Ras GDP/GTP exchange may exist. Indeed, such guanine nucleotide dissociation stimulators have been identified, first in *S.cerevisiae* as a product of the *cdc25* gene and later in *S.pombe* (Ste6) and *D.melanogaster* (SOS).[47,52,179]

While mammalian RasGDSs have been detected previously, only recently have they been identified and molecularly cloned. Mammalian homologs of CDC25 (mCD25 or CD25[Mm]) and SOS (mSOS) were first isolated from rat, mouse or human brain libraries.[241,344,397]

Other representatives of guanine nucleotide dissociation stimulators include Vav and Dbl proteins. Vav does not share sequence similarity with the yeast CDC25 protein, but instead contains a ca. 250 amino acid domain which is shared by members of the Dbl homology protein family. Since Dbl has been found to function as a GDS for a member of the Rho family of Ras-related proteins, it has been assumed that the homologous protein Vav would also function as a Rho family GDS.[153,199]

A distinctly different type of Ras GDS, designated smgGDS, has been described.[192] smgGDS contrasts with the CDC25-related GDSs in its substrate specificity. Furthermore, while the CDC25 homologous of RasGDSs can stimulate unmodified Ras proteins, Ras lipidation by isoprenylation is required for smgGDS stimulation.[266]

GUANINE NUCLEOTIDE DISSOCIATION INHIBITORS

A third class of regulatory proteins that control the Ras GDP/GTP cycle includes guanine nucleotide dissociation inhibitors (RasGDIs). RasGDIs act as negative regulators of Ras activity by virtue of their potent ability to inhibit dissociation of bound GDP.

Functionally related GDIs, which are active on members of the Ras-related Rab (RabGDI) and Rho (RhoGDI) proteins, have been molecularly cloned and characterized.[10,125] While RasGDIs recognize unmodified Ras proteins, the Rab and Rho GDIs interact only with prenylated proteins. Since prenylation is a stable and irreversible modification of proteins it has been speculated that these GDIs may mask the hydrophobicity of attached lipid groups to allow dissociation of Rho and Rab from membranes.[199] Thus these GDIs may play an important role in promoting the release of Rab and Rho from membranes. This role is consistent with the apparent need for Rab and Rho proteins to cycle between membrane and cytosolic compartments. Such translocations promote their functions in regulating intracellular vesicular transport and actin cytoskeletal organization respectively.

MECHANISM OF REGULATION OF RAS PROTEINS

The rapid and transient elevation of Ras-GTP levels triggered by various extracellular stimuli is believed to occur via either inhibition of GAP or GDI activity or stimulation of GDS activity. For example, GAP downregulation has been observed during phorbol ester induced T cell activation and in erythropoietin treated erythropoietic cells.[105] In contrast, both NGF-stimulated PC12 cells and EGF- or insulin-stimulated fibroblasts may exhibit elevation in Ras-GTP due to enhanced GDS activity.[250] Since both types of regulation have been observed, the mode of regulation appears to be specific for different cell types as well as for different stimuli.

How GAP activity may be downregulated is presently not clear. Experimental data indicate that p120GAP could be complexed with and phosphorylated by a number of ligand-activated receptor tyrosine kinases. It is suggested that p120GAP may provide a direct link between tyrosine kinases and Ras.

An alternative mechanism by which activated tyrosine kinases may mediate p120GAP activity is via the modification of some accessory proteins. An example of such a mechanism is the presence of two GAP-associated phosphoproteins p62 and p190.[70] Both proteins appear to represent substrates for phosphorylation by tyrosine kinases and are able to form complexes with N-terminal SH2 domains of p120GAP. This association has been found to reduce its GTPase stimulatory activity.[267]

Another mechanism by which GAP activity may be regulated during signaling is based on the observation that certain mitogenically stimulated lipids (arachidonic acid, phosphatidylinositol diphosphate and phosphatidic acid) may interact with p120GAP and NF1GAP and alter their abilities to stimulate RasGTPase function.[384] In general, while acidic lipids are inhibitors, some prostaglandins (PGD_{2a}, PGA_2) are stimulators of p120GAP activity.[157] This data has been obtained primarily by investigating in vitro systems; therefore it is not clear whether such lipids do indeed influence GAP activities under physiological conditions.

Recent studies have provided evidence for at least two mechanisms for upregulating RasGDS activities during extracellular stimulation. First, there is evidence that posttranslational modification of RasGDSs may cause activation of their catalytic function.[150] For example, the tyrosine phosphorylation of Vav during T cell activation correlates with its increased exchange activity. A similar phosphorylation of mSOS has been observed during EGF stimulation. Second, a series of recent reports have implicated proteins that contain SH2 and SH3 domains as critical components that directly couple activated tyrosine kinases with Ras activation by triggering the translocation of GDS to the plasma membrane where stimulation of Ras then occurs. One example of this kind of regulatory protein is the Grb2 molecule, consisting of SH2 and SH3 modules. Grb2 is believed to exist as a complex with SOS by virtue of its two SH3 domains, which recognize proline rich SH3 binding motifs present in the C-termini of the GDS protein. Upon ligand stimulation, the Grb2/SOS complex then associates with the autophosphorylated receptor tyrosine kinase via SH2 domain and consequently recruits Grb2-associated SOS to the plasma membrane where it can then stimulate Ras-GTP formation.[244] In addition to Grb2, a second SH2-containing protein, designated Shc (Src homology and collagen), has also been implicated in linking activated tyrosine kinases with Ras.[300] The results of several studies show that Shc is tyrosine phosphorylated transiently during activation of receptor tyrosine kinases and constitutively in cells transformed by nonreceptor tyrosine kinases, and this in turn promotes its association with Grb2 and SOS.[110,321] Thus, both Grb2 and Shc association with activated RTKs increases the concentration of SOS at the plasma membrane to promote activation of Ras proteins.

The above examples of Ras regulation obviously demonstrate only some of possible means for stimulation or inhibition of Ras functions. It is clear that some additional mechanisms of regulation of RasGDSs may exist. For instance, mCD25 appears not to have SH3 binding sequences. This observation suggests therefore that the activation in that case apparently occurs via a Grb2 independent pathway.

SUBFAMILIES OF RAS PROTEINS

The Ras superfamily of small GTP-binding proteins consists of several subfamilies, including in addition to Ras, the Rab, Rac, Rad, Rag, Ral, Ram, Rap and Rho proteins. The number of known Ras-related proteins tends to grow rapidly. Similarly, the biochemical pathways in which these proteins are known to actively participate become more and more diverse.

The Rho family constitutes one of the major branches of the Ras superfamily. Rho proteins share about 30% amino acid homology with Ras proteins and about 50% identity among the members of the Rho family.[191] The obtained experimental data indicate the importance of Rho proteins in, among other things, complex cytoskeletal rearrangements. There is emerging evidence that Rho family proteins may be downstream elements in the Ras signal transduction pathway.[316] The cross-talk between Ras and Rho family proteins can be accomplished by specialized GAPs and GDSs (e.g., mCD25 and p190) with substrate specificity to both types of low molecular weight proteins.[341]

The representatives of another family of Ras-related proteins are Rab proteins. At least 30 different Rab proteins have been identified that are specifically localized on different organelles and transport vesicles. These small GTP-binding proteins have emerged as major candidates for controlling the fusion events during regulated exocytosis and conferring specificity to membrane targeting and recognition.[394] Key elements in this process appear to be Rab regulatory proteins—GDS, GDI and GAP. Rab proteins remain membrane-associated under non-stimulated conditions. This membrane binding appears to be mediated in part by posttranslationally added geranylgeranyl moieties that are attached to cysteine residues at the carboxyl terminus. RabGDI specifically interacts with Rab proteins in the membrane-associated GDP-bound form, masks its hydrophobicity and allows removal of GDP-bound Rab from the membrane by means of forming a stable soluble complex. After removal from the membrane, Rab is thought to exchange its GDP for GTP. This reaction is speculated to be accomplished by specific guanine nucleotide dissociation stimulator. Rab proteins, for their normal functioning, should cycle between the GTP and GDP-bound form. Thus in order to activate GTP hydrolysis, they must interact with a GTPase-activating protein. In membrane docking and fusion, GAP may be localized on the plasma membrane to ensure the specificity of the membrane interaction, probably in concert with other proteins.

Recent advances in understanding of the control of NADPH oxidase activity in phagocytic cells helped define Rac members of the Ras superfamily of GTP-binding proteins as critical components in bactericidal reactions of phagocytes.[372]

NADPH oxidase removes electrons from NADPH to produce O_2^- from O_2 and can be rapidly activated by a variety of stimuli associated with leukocyte chemotaxis and phagocytosis (e.g., complement component C5a and formyl peptides produced as byproducts of bacterial protein secretion). This activation comes after the ligands binding to specific cell surface receptors present on neutrophils, macrophages and eosinophils. The NADPH oxidase consists minimally of four proteins. Associated with the plasma membrane is cytochrome b558, consisting of a 91 kDa glycoprotein subunit tightly bound to a smaller 22 kDa subunit. This cytochrome contains both flavin- and NADPH-binding sites and serves as the final electron donor for O_2. Two other oxidase components, p47*phox* and p67*phox*, are normally found in the cytosolic fraction of non-activated neutrophils and contain two SH3 motifs necessary for their interaction with other components of NADPH oxidase complex. The fourth protein required for a functional NADPH oxidase is Rac protein, defined as Rac2 in human neutrophils and Rac1 in guinea pig macrophages.[42] The latter protein must be isoprenylated for normal activity.

Rac is normally present in resting neutrophils as a cytosolic complex with GDI, but upon cell oxidative burst activation it releases from this complex and becomes associated with the plasma membrane and/or the cytoskeleton. The translocation process appears to be controlled by the conversion of Rac-GDP to Rac-GTP. This nucleotide dissociation is catalyzed by a membrane-bound GDS factor that may be activated during phagocytosis.

The kinetics of cytosolic factor translocation and NADPH oxidase activity suggest that oxidase function is dependent upon the coordinated translocation of p47*phox*, p67*phox* and Rac. It is also known that phosphorylation reactions are crucial for normal NADPH oxidase activation, as illustrated by the dramatic effects on O_2^- formation accomplished by addition of inhibitors of protein kinases and phosphatases. Thus one of functions of low molecular weight GTP-binding proteins can be to promote regulatory phosphorylations of NADPH oxidase complex representatives. In this connection it is of interest that one target for activated Rac has been identified as a protein kinase termed PAK-1.[42] The other proposed function of Rac may be the control of the ability of assembled system to transfer electrons to O_2. This function is speculated to be accomplished by direct interaction of Rac with both the cytochrome b558 and p67*phox* components.

=========CHAPTER 4=========

ADENYLYL CYCLASES

Some transmembrane mechanisms involved in the transformation of external signal into specific cell response involve the rapid production of intracellular second messengers. These second messengers relay the initial signal in a sequence of events acting to eventually regulate a specific physiological cell process at long term. This concept of signaling was initiated by the discovery of the role of cyclic AMP in numerous cellular processes and emphasized the role of the cAMP-generating enzyme, adenylyl cyclase, in cellular signaling.

DISTRIBUTION OF ADENYLYL CYCLASES

From earlier studies on the regulation of adenylyl cyclase it has been known that there are at least three adenylyl cyclase forms in mammalian tissues: the calmodulin-stimulated form in brain, the calmodulin- and G_s-insensitive form in testes, and the ubiquitously distributed G_s-stimulated form.[187] More detailed studies however indicate that the number of adenylyl cyclase types is much greater. Sequence homology analysis suggests that there are at least eight distinct types of mammalian adenylyl cyclases.[92] Additional members of these families may be identified and further families could be defined as additional tissues are examined. Comparison of the amino acid sequences indicates that the overall similarity between the various mammalian adenylyl cyclases is about 50%. However several individual sequences have considerably higher homology with each other.

The distribution of different adenylyl cyclase types is broad. Adenylyl cyclases of type 1 are neural-specific. In contrast, the Ca^{2+}-insensitive type 2 adenylyl cyclase mRNA is expressed at high levels not only in rat brain, but also in the olfactory bulb and olfactory epithelium. Type 3 mRNA was first reported to be restricted to the olfactory epithelium. Recent Northern blot analysis also detected type 3 mRNA in brain, spinal cord, adrenal tissues. Type 4 adenylyl cyclase has a wide tissue distribution and appears in brain and also in peripheral tissues such as heart, kidney, liver and lung but not in testis. The closely related types 5 and 6 of adenylyl cyclase are expressed in heart and neural tissue. Several other tissues, including kidney, liver, testes and skeletal muscle, also express type 5 mRNA. Type 7 adenylyl cyclase is widely expressed in brain, heart, kidney, liver, testes and skeletal muscle. Type 8 adenylyl cyclase is most abundant in brain, in particular cauda, cerebellum and hippocampus.[92]

STRUCTURE OF ADENYLYL CYCLASES

Adenylyl cyclases are integral membrane glycoproteins of relatively large size (1,080-1,248 amino acid residues).[93] The native purified proteins are 110-180 kDa. From analysis of the sequences, all the mammalian adenylyl cyclases are predicted to be transmembrane proteins with a complex topological structure of 12 transmembrane spans in two cassettes of six spans each (Fig. 10). Both the N- and C-termini are predicted to be cytoplasmic. It has been speculated that an important function of the transmembrane loops may be to provide optimal molecular interactions between the two cytoplasmic domains for efficient catalysis.

Two large cytoplasmic domains (a 350-amino acid loop between the first and second set of transmembrane domains and a 250-300 amino acid tail after the second membrane domain) were predicted to contain the catalytic core of the enzyme based on sequence similarity with cloned guanylyl cyclases. These cytosolic regions include also putative ATP-binding domains. Two intracellular domains share similarity with each other and both are highly conserved among the known mammalian adenylyl cyclase sequences. Expression of either half of the molecule does not yield a catalytically active enzyme. However when the two halves are coexpressed, substantial catalytic activity is restored. These data indicate that the catalytic domains act synergistically and expression of both is necessary for enzymatic activity.[93]

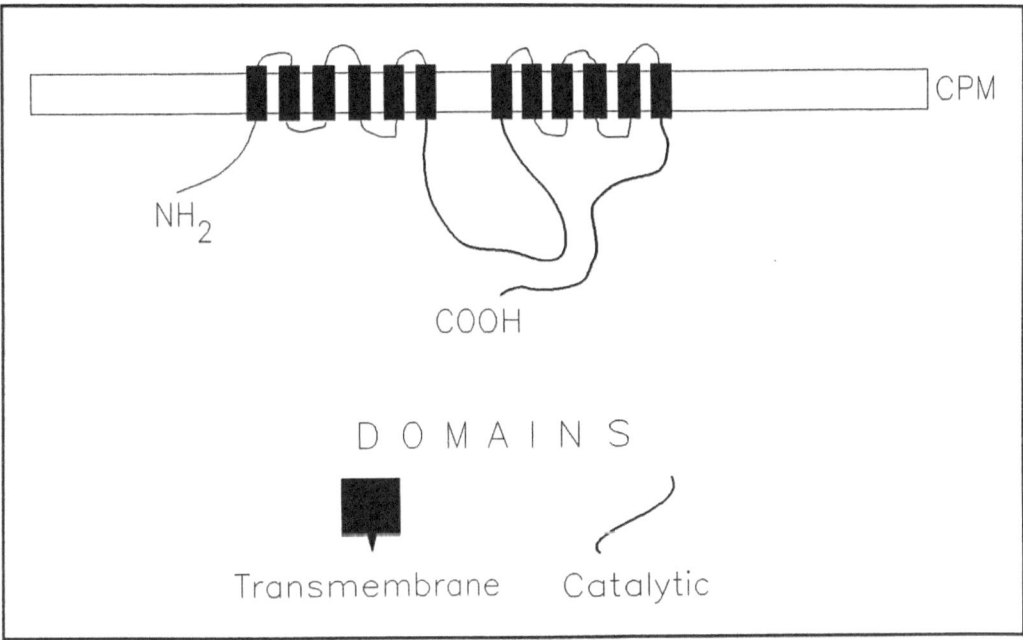

Fig. 10. Structural organization of adenylyl cyclases. CPM, cytoplasmic membrane. Both N- and C-termini of the enzyme are located in the cytoplasm of a cell. For details see text.

In addition to these regions there are non-overlapping domains which confer interaction with particular G proteins. Accumulated data indicate that these regions allow separate interaction with α_s, α_i and $\beta\gamma$ subunits of G proteins. Another functional domain is a sequence spanning amino acids 425-444 in the first cytoplasmic loop which is capable of inhibiting adenylyl cyclase activity. It could be that this sequence is an internal autoinhibitory domain. The site of Ca^{2+}/calmodulin binding in Ca^{2+}-stimulated adenylyl cyclases is believed to be in the first cytoplasmic loop, adjacent to the plasma membrane.

REGULATION OF ADENYLYL CYCLASES

Current cloning and expression experiments suggest that most if not all adenylyl cyclases are multiply regulated. The traditional view that $G\alpha$ subunits possess the dominant regulatory influence on adenylyl cyclase is being superseded by the fact that protein kinase C, Ca^{2+} and $\beta\gamma$ subunits of G proteins can stimulate or inhibit particular adenylyl cyclases far more effectively than the G protein α subunits. For example, based on studies of cells transfected with various adenylyl cyclases, the following observations have been made: phorbol esters elicit twice the stimulation of transfected type 2 adenylyl cyclase than does α_s; receptors operating through α_s barely stimulate type 1 or type 8 adenylyl cyclases, whereas Ca^{2+} stimulates their activity up to fourfold; type 6 adenylyl cyclase can be at least as effectively inhibited by elevation in Ca^{2+} as it can by G_i-linked receptors.

In general, adenylyl cyclases 1, 3 and 8 can be stimulated by Ca^{2+}, slightly stimulated by G_s and are unaffected by protein kinase C. $G\beta\gamma$ produces minor effects on these types of adenylyl cyclase. Adenylyl cyclases of 2, 4 and 7 types are insensitive to Ca^{2+}, but can be stimulated by α or/and $\beta\gamma$ subunits of G proteins. The effects of protein kinase C were reported as controversial. Adenylyl cyclases of types 5 and 6 are inhibited by Ca^{2+}, stimulated by G_s with no effect by $\beta\gamma$ or protein kinase C treatment. Within these groupings there is a range of responses; for instance, although both types 2 and 7 are stimulated by protein kinase C and G_s, type 2 is less susceptible to stimulation by G_s than by protein kinase C, whereas the opposite is true for type 7. In fact, of the eight adenylyl cyclase forms known to exist, no two are regulated in precisely the same manner.[93]

Another mechanism of regulation of enzymatic activity of adenylyl cyclases may be phosphorylation by cAMP-dependent protein kinase. Examination of the sequences of the cloned adenylyl cyclases showed that most of the enzymes contain one or two putative protein kinase A phosphorylation sites. The positions of putative protein kinase A phosphorylation sites usually were not conserved among the various adenylyl cyclases.[187] The possible influence of protein kinase A on adenylyl cyclase activities was confirmed later in biochemical experiments. In these studies treatment of membranes with protein kinase A resulted in decreased adenylyl cyclase activity.[307] This kind of modification may represent a mechanism of feedback regulation and exemplifies one component of desensitization of adenylyl cyclase response. The varying location of the protein kinase A sites suggest that different types of adenylyl cyclases may be differentially regulated by cAMP-dependent phosphorylations.

CHAPTER 5

PHOSPHOLIPASES

In addition to production of cAMP by adenylyl cyclases, hydrolysis of phospholipids has been shown to be another major mechanism in the generation of potent second messengers during intracellular signaling processes. Among the best studied precursors of lipid signaling molecules are phosphatidylinositol (PI) and phosphatidylinositol-phosphates (PIP$_n$), phosphatidylcholine (PC), lysophosphatidylcholines (lysoPC) and sphingomyelin (SM). These precursors can be hydrolyzed by specific phospholipases to produce inositol phosphates, diradylglycerols (DRGs), ceramide and free fatty acids (FFAs). In addition, certain intermediates of phospholipid breakdown (e.g., lysophospholipids or phosphatidic acids) are also shown to be potent messenger molecules.

The various cellular phospholipases (PLs) participating in signaling are classified according to the bond cleaved in a phospholipid. Thus phospholipases A_1 and A_2 selectively remove fatty acids from the sn-1 and sn-2 positions. Phospholipase B is a lysophospholipase which can cleave both the 1- and 2-lysophospholipids produced by PLA_2 and PLA_1 respectively. Phospholipase C cleaves the bond between glycerol and phosphate, phospholipase D splits off the amino alcohol moiety from a phospholipid, whereas sphingomyelinase produces ceramide and phosphocholine from sphingomyelin (Fig. 11). The other main enzymes with possible signaling effect include diacylglycerol (DAG)- and monoacylglycerol (MAG)-lipases, ceramidase and phosphatidic acid (PA) phosphohydrolase.[148]

The broadest range of regulating activities is possessed by phospholipases C, especially phosphatidylinositol-specific phospholipase C (PI-PLC). The action of the latter enzyme produces two potent

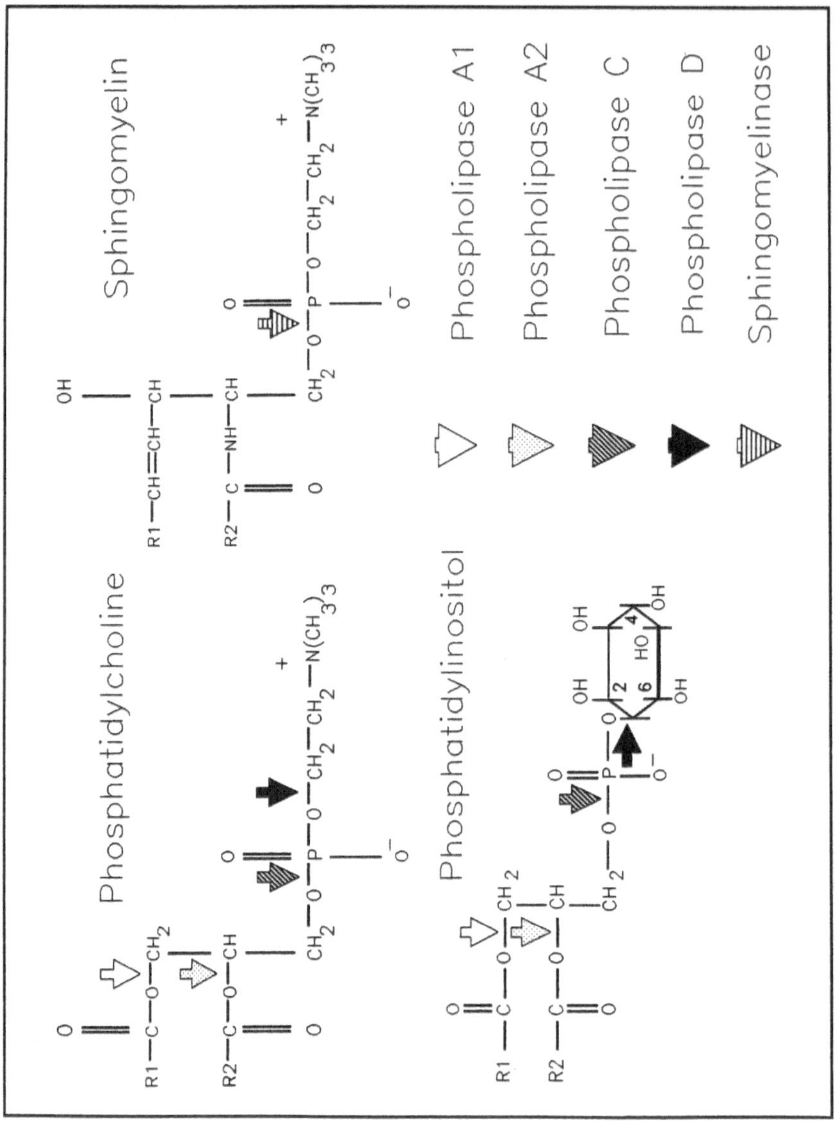

Fig. 11. Main classes of phospholipases participating in signal transduction in eukaryotes. Sites of
cleavage in target phospholipids
are shown by arrows.

second messengers—diacylglycerol and inositol-1,4,5-triphosphate (IP$_3$). One of the most physiologically important effects of liberated diacylglycerol is activation of protein kinase C. The primary effect of IP$_3$ is an increase in intracellular Ca^{2+} level.

LIBERATION OF DIACYLGLYCEROL

The most prominent routes with respect to the involvement of DAG in signal transduction have been shown to originate from the hydrolysis of the three phosphoinositides—phosphatidylinositol, phosphatidylinositol-4-phosphate and phosphatidylinositol-4,5-diphosphate—by phosphatidylinositol-specific phospholipase C.[155]

There are multiple PI-PLC enzymes in mammalian cells as deduced from direct protein isolation and molecular cloning studies. At least ten isoforms of PI-PLC have been described, varying from 70 to 154 kDa in size, which are categorized into three groups designated β, γ and δ. In addition to molecular mass, these isoforms were shown to differ in pI, pH optimum, calcium dependence, cleavage properties and mode of activation.

All representatives of phosphatidylinositol-specific phospholipases C share two highly conserved domains that form the catalytic region. Since substrates of PI-PLC usually are localized in the membrane bilayer, enzymes to accomplish hydrolysis of phosphatidylinositols must bind to membranes. In many instances SH2 and SH3 domains of phospholipases play a critical role in such translocations.

Different PI-PLC isozymes are regulated by different mechanisms. All PLC-β isozymes are activated by G proteins. The degree of this regulation may be broad. Thus G$_i$ may cause weak stimulation of PLC activity in contrast to strong stimulation by G$_q$ class of heterotrimeric GTP-binding proteins. PLC-γ is activated by direct phosphorylation on tyrosine residues. Its stimulation by the EGF receptor, for example, requires that PLC-γ first binds to the C-terminal autophosphorylation domain of the EGF receptor. This binding is mediated by the SH2 domains of PLC-γ and results in phosphorylation of PLC-γ on tyrosine residues. How PLC-δ is activated remains unknown, although it was found to be stimulated by FFAs at micromolar Ca^{2+} concentrations.[231] In addition to PLC-γ, phosphorylation of which is the main route of activation, all members of PI-PLCs possess consensus sequences for phosphorylation by different types of protein kinases. Thus phosphorylation and dephosphorylation could be modes of fine regulation of these enzymes. The other possible means of activation of PI-PLC in general is by supplying calcium as a cofactor.

There is now increasing evidence that DAG with messenger action is also generated from the hydrolysis of phosphatidylcholine by phosphatidylcholine-specific phospholipases C (PC-PLCs). The first indication came from measurements of DAG accumulation in response to vasopressin in rat hepatocytes.[40] These data were later confirmed by applying different agonists in various cellular systems. It is also becoming clear that many agonists such as growth factors, hormones and neurotransmitters stimulate PC breakdown in addition to phosphatidylinositol hydrolysis.

The regulation of PC-PLC activity is accomplished in many ways analogously to that of PI-phospholipases. The stimulation of phosphatidylcholine breakdown is accomplished primarily by specific G proteins. The other important pathways of PC-PLC regulation is via protein kinase C action and by alterations in intracellular Ca^{2+} contents.

The physiological significance of phosphatidylcholine hydrolysis may be based on the prolonged formation of DAG, which is important in cellular mechanisms that require long-term activation of protein kinase C. Thus PC hydrolysis can act to sustain a message that was initially transmitted via inositide breakdown. It is possible that the initial transient increase of DAG arising from PIP_2 may act as a trigger for PC hydrolysis through activation of protein kinase C. In concordance with this hypothesis a biphasic elevation of cellular DAG was observed comprising a rapid initial transient diacylglycerol elevation due to the hydrolysis of phosphoinositides and a second rise to a plateau phase as a consequence of PC breakdown.

However, in addition to these data there is strong evidence indicating that the products of phosphatidylcholine hydrolysis could affect signaling pathways different from those regulated by PI breakdown. The predominant components of DRG in tissues are the ester-linked 1,2-diacyl species (DAG) and the 1-alkyl-2-acyl-glycerols (AAG). However it is known that additional forms of PC exist and are made up of different molecular species, including ester-, ether- and vinyl-ether-linked species. Further diversity exists, as the acyl chain structure can vary greatly. It is possible therefore that specific DRGs generated from PC are recognized by special isotypes of protein kinase C that act in a different domain than does PIP_2-stimulated protein kinase C, and this may provide an additional mechanism to maintain signal specificity.[410]

The third pathway by which DAG can be formed is by a concerted action of two hydrolytic enzymes involving phospholipase D (PLD) and a phosphatidic acid phosphohydrolase. In this pathway phosphatidylcholine is first hydrolyzed by PLD to phosphatidic acid which is then degraded to DAG by PA hydrolase.[34]

The receptor-mediated regulation of PLD activity is in many details similar to that of PLC. The main components of such regulation include protein kinase C, Ca^{2+} and G proteins. In addition to these modes of PLD activation, several studies suggest a role for tyrosine phosphorylation in receptor-linked activation of phospholipase D. For instance, it has been shown that EGF stimulation of fibroblast PLD is inhibited by specific inhibitors of protein tyrosine phosphorylation but not protein kinase C inhibitors.[91]

There is increasing evidence to suggest the specific involving of phospholipase D in many antibacterial cellular responses including phagocytosis, respiratory burst etc. Phagocytosis by human neutrophils mediated by complement receptors has been shown to be associated with phosphorylation of a protein kinase C substrate (MARCKS protein) and DAG generation.[116] Since under these conditions PIP_2 was not hydrolyzed, these observations suggest that phosphatidylcholine-derived DAG is capable of activating neutrophil protein kinase C with subsequent stimulation of phagocytosis. Use of a series of inhibitors has led to the conclusion that DAG involved in the initiation and maintenance of the respiratory burst in human neutrophils has been produced from PC by phospholipase D.[301]

In other studies using human neutrophils as a model, TNF-α was shown to augment superoxide anion production induced by fMLP and this augmentation correlated well with PA generation through PLD. In rabbit neutrophils ethanol inhibited both the release of granule enzymes and accumulation of PA and DAG.[195]

Propranolol, which inhibited DAG accumulation while enhancing PA level, did not inhibit enzyme release, thus implicating a role for PLD-derived PA in neutrophil degranulation.[18]

The mechanism of phosphatidic acid action is currently not clear. One proposition exists that PA could accomplish its effect by altering the activity of low molecular weight GTP-binding proteins. It was demonstrated that PA inhibited the activity of the GTPase activating protein, thus increasing the activity of cellular Ras. Members of the Ras superfamily in turn are believed to be critical in the regulation of different cellular processes linked to phagocytosis, including cytoskeleton rearrangements and NADPH oxidase activity.[383,384]

Available evidence indicates that phosphatidylcholine is the preferred substrate for receptor-stimulated PLD. However in some systems such as NIH 3T3 fibroblasts, PLD was shown to catalyze hydrolysis of both PC and phosphatidylethanolamine following stimulation with growth factors.[204] A recent study has provided evidence for PLD-catalyzed hydrolysis of phosphatidylinositol in bradikinin-stimulated Madine-Darby canine kidney cells.[177] Thus, a widespread degradation by phospholipase D of phospholipids other than PC, in particular phosphatidylethanolamine and phosphatidylinositol, is becoming increasingly evident.

LIBERATION OF INOSITOL-PHOSPHATES

In 1983 Michael Berridge demonstrated that D-*myo*-inositol 1,4,5-triphosphate (IP$_3$) generated in the course of receptor-activated inositol lipid turnover was a signaling second messenger for release of stored Ca^{2+} from intracellular sites. Subsequent experiments demonstrated that receptor-activated Ca^{2+} mobilization via the inositol phosphate signaling system usually involves two phases: (1) rapid Ca^{2+} release from an intracellular store, and (2) a more prolonged phase of Ca^{2+} entry from the extracellular space.[29,309]

The first phase of Ca^{2+} mobilization begins with the stimulation of plasma membrane receptors which activate phosphatidylinositol-specific phospholipase C. The latter enzyme generates IP$_3$ from plasma membrane phosphatidylinositol-4,5 diphosphate. The liberated IP$_3$ then diffuses to specific receptors which are ligand-activated calcium-selective channels present on an intracellular calcium-storing organelle. The binding of IP_3 increases the probability of channel opening, which allows Ca^{2+} to flow into the cytoplasm.

The second phase of the Ca^{2+} signal likely does not result from the direct action of either a plasma membrane receptor or inositol phosphates but appears to operate through a "capacitative" mechanism.[308] Following the proposed hypothesis the empty calcium-storing structures produce a retrograde signal that activates Ca^{2+} influx across the plasma membrane. The nature of such a signal is unknown to date and was proposed to include participation of metabolites of cytochrome P-450, cyclic GMP, $(1,3,4,5)IP_4$ or some novel diffusable messenger. It is likely also that Ca^{2+} itself could activate Ca^{2+}-permeable channels in the plasma membrane either directly or through Ca^{2+}-induced conformational changes in the actin cytoskeleton. In the latter case the conformational changes in the cytoskeleton could be conveyed to the plasma membrane Ca^{2+} channels through attachments with plasma membrane actin-binding proteins. Reports from several laboratories suggest also that phosphorylation of several proteins on tyrosine residues may be important in the pathway of Ca^{2+} entry activation.[309]

The subsequent metabolism of inositol phosphates is complex. Before the discovery of the Ca^{2+}-mobilizing actions of $(1,4,5)IP_3$, Downes and colleagues[103] demonstrated the presence of specific enzymes that remove the phosphates from $(1,4,5)IP_3$, ultimately generating free inositol. This pathway at first appeared to function as a terminator of the Ca^{2+} signal and to contribute to recycling of inositol for the restoration of precursor inositol lipids. The first indication that inositol phosphate metabolism was more complex than originally realized was represented by the discovery of an inositol phosphate isomer, $(1,3,4)IP_3$, formed on prolonged receptor activation. Subsequently it was demonstrated that this IP_3 isomer was formed by sequential phosphorylation of $(1,4,5)IP_3$ on the 3-position by 3-kinase, followed by dephosphorylation at the 5-position by 5-phosphatase. The majority of $(1,3,4)IP_3$ was further dephosphorylated eventually to inositol, but $(1,3,4)IP_3$ was also phosphorylated to form $(1,3,4,6)IP_4$, $(1,3,4,5,6)IP_5$ and possibly also IP_6. Menniti and colleagues[255] demonstrated that another inositol tetrakisphosphate isomer $(3,4,5,6)IP_4$ was derived from $(1,3,4,6)IP_4$

and its formation was markedly increased by PLC-linked agonists. It appears likely that this reaction provides some function in cell regulation although its importance remains a mystery. An additional complication in the inositol phosphate metabolic scheme was the demonstration of the cyclic inositol formation $(c1:2,4,5)IP_3$. This product was shown to be accumulated as a minor proportion of the outcome of PI-PLC action on phosphatidylinositol 4,5-diphosphate. The majority of experimental findings indicate that the described metabolites are inactive in mobilizing Ca^{2+} and their functions in cellular signaling remain to be elucidated.

Several second messenger-generating systems are known to antagonize the agonist-evoked Ca^{2+} elevations. Downregulation of the Ca^{2+} responses occurs notably by stimulation of protein serine/threonine kinases, especially protein kinase C, protein kinase A and protein kinase G. However other factors are likely to be involved as well.[163]

Inhibitory effects of protein kinase C on receptor-mediated IP_3 production and Ca^{2+} mobilization are widespread. Since protein kinase C is activated by DAG which is produced by PLC, this inhibition is considered as a negative feedback loop in the Ca^{2+} response. It has been proposed that this feedback inhibition at the level of protein kinase C may be responsible for the generation of sinusoidal Ca^{2+} oscillations.[309] Additionally, protein kinase C may stimulate the degradation of IP_3 by specific phosphatases.[202]

Elevated levels of cyclic nucleotides potentially downregulate the Ca^{2+} responses in platelets. These effects of cAMP and cGMP are most probably mediated by protein kinase A and protein kinase G which, at least in platelets, are expressed at relatively high levels. In agreement with this proposition several authors have found that cyclic nucleotides inhibited PLC-induced phospholipid hydrolysis and inositol phosphate production. In other experiments it was disclosed that cAMP-dependent phosphorylation of IP_3 receptors reduce Ca^{2+} release. Overall, the present literature suggests that several steps of Ca^{2+} signaling are influenced in parallel by cyclic nucleotide-dependent protein kinases, although the points of action of the kinases are still incompletely known.

LIBERATION OF FATTY ACIDS

In recent years evidence has been accumulating that in addition to the well-established second messengers the existence of additional messenger molecules should be taken into account. Several studies, carried out in various cell types, suggest that fatty acids (in particular unsaturated fatty acids) can behave as second messengers and modulators through their capacity to regulate the phospholipases, protein kinases, G proteins, adenylyl and guanylyl cyclase and ion channel activities as well as acting on other structures involved in stimulus-response mechanisms.

One of the well studied fatty acid signaling molecules is arachidonic acid. Arachidonic acid itself and its metabolites, i.e., prostaglandins, leukotrienes and tromboxanes (collectively referred to as eicosanoids) are potent regulators of various physiological responses. These metabolites govern cellular functions as diverse as inflammation, natural killer activities, smooth muscle contraction, ion channel activities, neurotransmission and visual signal transduction. Many of the effects accomplished by eicosanoids are directed through specific receptors on the surface of eukaryotic cells. However numerous pathways could be affected by direct action of intracellular arachidonic acid and its metabolites on certain signaling structures. Mammalian cells do not store eicosanoids and unstimulated cells contain very little, if any, non-esterified arachidonate. Therefore the levels of these compounds are primarily regulated by the availability of free arachidonic acid, which is supposed to be liberated mainly from phosphatidylcholine by phospholipase A_2. In addition, it also can be produced by the activation of PLC followed by DAG- and MAG-lipases; by PLD followed by phosphatidic acid phosphohydrolase, DAG- and MAG-lipases; and finally by PLA_1 followed by lysophospholipase PLB.

PLA_2, the main enzyme in the generation of arachidonic and other unsaturated acids, is activated by many agonists, including tyrosine kinase activators EGF and PDGF.[280] There is increasing evidence that heterotrimeric G proteins mediate the stimulation of PLA_2 in many cell types, including neutrophils, fibroblasts, sensory neurons and retinal rods. Based on the well-documented stimulation of adenylyl cyclases and phospholipases C by $G\alpha$ it was initially believed that the α subunit is the active component triggering PLA_2. Recently however a few examples of direct effects of $\beta\gamma$-subunits or single β subunits have been reported. An impressive body of information suggests also the regulation of cytosolic PLA_2 by Ca^{2+}. In addition certain isoforms of PLA_2 contain multiple phosphorylation sites for protein kinases among which are sites for a mitogen-activated protein kinase. It appears that this enzyme plays an important role in PLA_2 activation.

In addition to the above modes of PLA_2 regulation, several studies have shown that free fatty acids exert a modulatory effect on the activity of this enzyme as well as on PLC and PLD. It has been shown that the activity of a membrane-bound PLA_2 purified from a macrophage-like cell line was inhibited by arachidonic, oleic, linoleic, linolenic and eicosapentaenoic acids. Arachidonic acid was the most inhibitory fatty acid and behaved kinetically as a competitive inhibitor. On the other hand, the saturated fatty acid palmitate caused no enzyme inhibition.[232] These results suggest that fatty acid inhibition might represent a negative feedback regulation for PLA_2 activity.

The biological effects of eleborated free fatty acids are diverse. Casabiell and colleagues [62] reported that *cis*-unsaturated FFAs (oleic and palmioleic acids), but not *trans*-unsaturated or saturated FFAs, blocked the rapid rise in cytosolic calcium concentration in response to different stimuli. They suggested that *cis*-unsaturated FFAs blocked signal transduction by interfering with receptor-PLC interaction or eventually with PLC-substrate interaction. The effect of free fatty acids seemed to be dependent on their chemical configuration rather than hydrophobicity or chain length.

A large number of studies have shown that PLD activity was stimulated by fatty acids, in particular by oleic acid. Chalifour and Kanfer [66] reported that mono-unsaturated fatty acids were the most effective activators of rat brain microsomal phospholipase D; poly-unsaturated fatty acids stimulated it to a lesser degree, while saturated fatty acids were ineffective.

A number of studies have shown that certain free fatty acids may be second messenger molecules for the regulation of ion channels. This may involve direct interaction of the fatty acid with the ion channel itself (or an accessory protein) or alteration of the interaction of channels with the lipid bilayer. Wolf and colleagues[401] showed that in digitonin-permeabilized pancreatic islets, exogenous arachidonic acid elicited significant Ca^{2+} release from endoplasmic reticulum. This arachidonic acid-induced Ca^{2+} mobilization occurred rapidly, within 2 min, and was not due to the metabolites of arachidonic acid. One could hypothesize that arachidonic acid might act indirectly by inducing the PI-PLC activity. However recent reports showed that a direct action of arachidonic acid on Ca^{2+} mobilization not mediated by IP_3 might also exist. In this context it has been shown that arachidonic acid has no effect on polyphosphoinositide breakdown in neutrophils, although it does elevate cytosolic free Ca^{2+} concentrations in these cells.[393] Recently it was shown[178] that in guinea pig ventricular myocytes long chain unsaturated (oleic, linoleic, linolenic and arachidonic) fatty acids induced an increase in voltage-dependent calcium currents. In contrast, neither short chain fatty acids (<12 carbons), nor fatty acid esters (oleic and palmitic methyl esters) had an effect. On the whole, these observations suggest that arachidonic acid and other unsaturated fatty acids play the role of second messengers in transducing activation signals by inducing Ca^{2+} mobilization from intracellular stores or by modulating plasma membrane calcium channels.

Although some studies indicate that arachidonic acid might affect G proteins via its metabolites there is evidence that arachidonic acid itself is capable of directly modulating G protein-mediated signals. It is known that arachidonic acid induces polymorphonuclear leukocyte activation (cell aggregation, superoxide anion generation and release of lysosomal content). These effects could be accomplished by eicosanoids acting on the specific cell surface receptor. Recently however, the studies of Abramson and colleagues[1] on human neutrophils suggested that cell activation was resultant to the direct stimulation of a G protein by arachidonic acid and other *cis*-unsaturated fatty acids. The mechanism of such stimulation may be linked to an arachidonic acid-dependent increase in the number of available GTP binding sites on G protein.

An effect of FFAs on guanylyl and adenylyl cyclase activities has been demonstrated in many studies. Micromolar concentrations of both linoleic and linolenic acids enhanced adenylyl cyclase activity in murine neuroblastoma cells.[272] Louis and colleagues[234] showed that purified soluble guanylyl cyclase from rat brain was directly stimulated by arachidonic acid at concentrations ranging from 1 μM up to 500 μM, whereas higher concentrations were strongly inhibitory. Moreover there is evidence that both soluble and particulate guanylyl cyclases may be directly activated by arachidonic acid released from membrane phospholipid after incubation of whole cells or enzyme assay mixtures with PLA_2.

During several investigations the influence of arachidonic acid and its 12-lipooxygenase metabolite on calmodulin-dependent protein kinase 2 and cAMP-dependent protein kinase activities have been shown. The other example of the role of fatty acids as second messengers and modulators is the activation of protein kinase C by fatty acids.[216,352]

It is interesting to differentiate the experimental systems where fatty acids have been shown to be second messengers by directly activating protein kinase C in DAG-, phosphatidylserine- and Ca^{2+}-independent manners, and those systems in which fatty acids were modulatory molecules greatly enhancing the DAG-dependent activation of protein kinase C and allowing the enzyme to exhibit almost full activation at nearly basal, relatively low Ca^{2+} concentrations.

Murakami and colleagues[271] showed that unsaturated fatty acids, particularly oleic acid (25-400 μM), directly stimulated protein kinase C activity purified from rat brain in the absence of DAG and phosphatidylserine. Oleic acid activation of protein kinase C did not require Ca^{2+} and did not depend on the Ca^{2+} concentrations when Ca^{2+} contents were at or below 10 μM. In another study it has been shown that arachidonic acid induced phosphorylation of only a subset of the phosphoproteins which were shown to be produced by phorbol ester activation of protein kinase C.[198] Thus protein kinase C isozymes may show specific substrate preferences under physiological conditions in response to arachidonic acid or DAG.

The activation of protein kinase C in intact human platelets is an example of the role of fatty acids as modulators (enhancers) of the DAG and phorbol ester effects on protein kinase C. Several *cis*-unsaturated fatty acids at <50 μM concentrations added directly to intact human platelets greatly potentiated protein kinase C activation.[408] This enhancement absolutely required the presence of DAG or tumor-promoting phorbol ester. Saturated and *trans*-unsaturated fatty acids were found to be inactive. Thus free fatty acids may act as modulators to sustain protein kinase C activity when DAG concentrations drop. In contrast, the second messenger role of unsaturated fatty acids would be to substitute for DAG and to directly activate certain isoforms of protein kinase C. The importance of one or the other roles would depend on the tissue, cellular and intracellular localization of the different isoforms of protein kinase C.

SPHINGOMYELINASES

Sphingomyelin hydrolysis with production of ceramide and phosphocholine was recognized as a potential signaling pathway when Okazaki and colleagues showed that vitamin D3 stimulated a sphingomyelinase activity in HL-60 cells, thus leading to accumulation of ceramide. The hydrolysis of sphingomyelin by sphingomyelinase could be activated by multiple agonists, including 25-dihydroxyvitamin D_3, TNF-α and γ-interferon.[256]

Whereas phosphatidylcholine and phosphatidylinositol hydrolysis generate a series of messengers that sustain the protein kinase C activities, sphingomyelin is hydrolyzed to generate messengers capable of termination the phosphorylation cascade. Ceramide is a potent inhibitor of cell growth as well as a promoter of cell differentiation. Its metabolite, sphingosine, is a potent inhibitor of protein kinase C that acts by blocking DAG activation.[257] Another metabolite, lysosphingomyelin, is also formed during hydrolysis of sphingomyelin and also was shown to inhibit protein kinase C.[158]

GPI-ANCHOR-HYDROLYZING PHOSPHOLIPASES

The action of glycosylphosphatidylinositol-hydrolyzing phospholipases (GPI-PLs) may result in the generation of the second messengers diradylglycerol and phosphoinositol-glycan, or PA and inositol-glycan from GPI anchored to plasma membrane.

In eukaryotic organisms several GPI anchor-hydrolyzing phospholipases C have been identified in rat liver, mouse brain and in

T.cruzi.[51] In the case of rat liver GPI-PLCs were suggested to be a possible target of insulin action. It was postulated that insulin activated a GPI-PLC which cleaved a GPI lipid in the plasma membrane thereby releasing phosphoinositol-glycan mediators. An interesting observation has been done by comparison of actions of added insulin and its mediators. It was demonstrated that such phosphoinositol mediators mimicked many of the effects of insulin when applied to various intact cells and modified the activity of various enzymes in vitro in a manner consistent with the effects of insulin upon these enzymes in intact cells.[200]

It is unclear whether IG-mediators are generated by the hydrolysis of free or protein-linked glycosyl-PI. It has been reported that insulin stimulated the release from cell membranes of some GPI anchored proteins. If glycosyl-PI anchored proteins serve as precursors to IG-mediators then a proteinase activity in addition to PLC activity would be required for the release of free mediator.

Besides GPI-specific phospholipase C, the existence of GPI-PLD is well established in mammalian placenta, serum, brain, pancreas, liver etc. It is possible that GPI-PLD, analogously to phospholipase C, could be also involved in intracellular processing of GPI-anchored proteins.

=========== CHAPTER 6 ===========

NITRIC OXIDE
SYNTHASES

Nitric oxide (NO) is a biologically active product enzymatically formed from a terminal guanidino-nitrogen of L-arginine by members of the NO synthase (NOS) gene family. Members of NO synthases are classified into three isoforms NOS 1, 2, and 3. This original classification of NOS was based on the physical and biochemical characteristics of the purified enzymes, i.e., subcellular location (soluble vs. particulate fraction), regulation by free Ca^{2+} concentrations and inhibition by specific inhibitors.[336]

All NO synthases are homodimers of subunits with a range in size between 130 and 160 kDa. None of the described NO synthase isoforms contains an amino acid sequence suggestive of a transmembrane domain. The N-terminal half of all NOS contains binding motifs for NADPH, FAD and FMN. Additional binding sites within the NOS polypeptide were postulated for heme, tetrahydrobiopterin and L-arginine. The predicted amino acid sequence of NOS 1 has consensus sites for phosphorylation by cAMP-dependent protein kinases. However forskolin-induced increase in cAMP-dependent protein kinase activities and intracellular cAMP levels did not regulate NOS. In contrast, protein kinase C and Ca^{2+}/calmodulin-dependent protein kinase 2 phosphorylated NO synthases on both serine and threonine in vitro. These phosphorylations resulted in a marked decrease of NOS activity.

Other types of NOS regulation include posttranslational lipidation, such as myristoylation, and glycosylation. These modifications affect both the subcellular localization of the enzyme and its specific activity.[278]

All NO synthase forms bind calmodulin and have conserved consensus sequences for calmodulin binding. In the case of NOS 1 and 3 interaction with calmodulin depends on elevated intracellular free Ca^{2+} concentrations. At resting Ca^{2+} concentrations (<100 nM) these NO synthases are calmodulin-free and inactive. They bind calmodulin and become fully active at increased Ca^{2+} concentrations (>500 nM). Similar to NOS 1 and 3, NO synthase 2 is calmodulin-free and inactive at <100 nM of free Ca^{2+}. Its activity increases upon reconstitution with calmodulin, but unlike NOS 1 and 3 this activation is Ca^{2+}-independent. Thus, calmodulin binding represents a common activation principle of all NO synthases, whereas free intracellular Ca^{2+} affects their activity differentially.

The probable importance of NO depends on its activity as a paracrine substance and also as an intracellular messenger molecule. The short half life of NO argues against an important humoral role. The main if not sole effect of low NO concentrations is stimulation of soluble guanylyl cyclases. However high concentrations of NO which are generated by the immunologically induced NOS 2 can interact with almost every cellular protein. Under in vivo conditions, reactions with NO have been described for oxygen, oxygen radicals, metal-, iron-, heme- and thiol-containing proteins and primary amines.[356] In many cases these reactions lead to loss of function of the entire cell or particular enzymes. Activated macrophages, which possess a high-output NOS 2, utilize some of these reactions as non-specific immune defense mechanisms against bacterial, protozoan and viral infections.

CHAPTER 7

SOLUBLE GUANYLYL CYCLASES

In addition to phospholipases, NO synthases and adenylyl cyclases which synthesize lipid second messengers, NO and cAMP respectively, guanylyl cyclases are also capable of production of the potent transducer molecule cyclic GMP. Two forms of guanylyl cyclase exist a soluble one and a particulate or receptor form.[76]

A soluble form of guanylyl cyclase (sGC) is found in most mammalian tissues. The most rich source of the enzyme is lung. The lung guanylyl cyclase has a molecular mass of about 150 kDa and exists as a heterodimer of 82 kDa α and 70 kDa β subunits as determined by SDS-PAGE. The molecular cloning of the subunits revealed that their C-terminal regions were homologous with that of plasma membrane guanylyl cyclases; however there was virtually no similarity within N-terminal regions. Expression of single α or β subunits in mammalian cells resulted in no detectable GC activity. The coexpression of both subunits, however, resulted in sodium nitroprusside-stimulatable GC. Therefore, although each subunit contains an apparent catalytic domain, they may not function independently. Alternatively, only one of the soluble enzyme subunits may contain catalytic activity, but it may require the second subunit as an activator (or deinhibitor).[339]

It is likely that the folding of two monomeric sites represents the minimal necessary structure of an active GC. It is possible also that functionally active cyclases exist as tetramers or higher-ordered structures. In such a case the potential diversity in signaling properties of the cyclases could be significantly expanded in the face of relatively few enzyme genes.[137]

Four different cDNA clones (α1, α2, β1, β2) encoding homologous proteins have been obtained for subunits of the soluble GC,[135,211,409] but only the heterodimer α1β1 has been convincingly demonstrated in tissue.[134,141,180] In addition, cloning experiments showed that coexpression of α2β1 resulted in the formation of an enzyme with catalytic activity, but it is not known if α2β1 exists normally in vivo. Whether or not soluble homodimers are also formed is not yet known. If they exist, the homodimers of α1 or β1 apparently possess low enzyme activity relative to that of the heterodimer.

Purified soluble guanylyl cyclase (α1β1) contains iron and copper as transition metals. In addition, two putative heme binding domains have been suggested for sGC: (1) a conserved cysteine which is present both in the α and β subunits and has flanking sequences that are identical to those which coordinate heme in fish cytochrome P_{450}, and (2) a histidine residue which is conserved in the β subunit of sGC. The heme-binding motifs are absolutely necessary for proper activity of sGC and it is the heme that has been proposed to act as the receptor for nitric oxide and carbon monoxide.[141,180,391] Thus heme-free guanylyl cyclase no longer responds to NO although basal sGC activity could be detected in those preparations.

The mechanism of sGC activation is not known in full detail. The proposed chains of events are as follow: NO binds to the sixth coordination position of the heme iron moiety of sGC, dislocates the heme-iron and finally induces a conformational change in sGC. The latter step either de-inhibits or activates the catalytic site of sGC, stimulating its activity by as much as 170-fold.[180,382]

In addition to NO, soluble guanylyl cyclases are shown to be regulated by enzymatically formed activating factors that include CO and OH. Although NO synthases have probably the widest distribution in the body and have a clearly established role in sGC regulation, other enzymes, e.g., CO-generating heme oxygenase, colocalize better with sGC in the central nervous system than do NOS.[56,239,391,411]

CO could be generated from at least two precursors, fatty acids and heme. In the former case polyunsaturated fatty acids should be subjected to NADPH-dependent enzymatic peroxidation.

Smaller amounts of CO could arise from cytochrome P_{450}-catalyzed microsomal lipid peroxidation. An alternative pathway for CO generation could include oxidative heme destruction. This process is provided primarily by heme oxygenases which convert iron protoporphirine IX and several other hemoproteins to biliverdin IXα and CO. Among minor routes is P_{450}-catalyzed iron protoporphirine IX destruction. One of the possible enzymatic pathways of OH production utilizes xanthine oxidase as a key enzyme.

The mechanism by which CO activates sGC is probably identical to that of NO since both CO and NO have similar effects on platelet aggregation, vascular smooth muscle function and intracellular cGMP levels. The mechanism of activation of soluble guanylyl cyclase by OH is not fully known but appears to be different from that of NO and CO. One possible mode of regulation is interaction of OH with regulatory thiol groups of sGC which function as cellular redox sensors.

In some contrast to cAMP, which performs its action primarily through cAMP dependent protein kinase, cGMP produced by guanylyl cyclases has several intracellular targets. The main targets of cGMP action include cGMP-dependent protein kinases and cGMP-gated channels. However other signaling enzymes could be regulated by this molecule as well, including phosphodiesterases, cAMP-dependent protein kinases and ADP-ribosyl cyclases.

CHAPTER 8

PROTEIN KINASES

Of the many types of posttranslational modification of proteins that occur in cells, relatively few are readily reversible. The most common type of reversible protein modification is phosphorylation. It has become clear that nature has chosen phosphorylation-dephosphorylation as an almost universal mechanism for regulating the functions of proteins, not only those that display enzymatic activity but also proteins involved in many other biological processes. This importance of protein kinases in regulating cellular activities is underscored by the large number of protein kinase genes that are present in eukaryotic genomes. The rough estimate is that humans have as many as 2000 conventional protein kinase genes.[181]

The protein kinase reaction is dependent on a nucleoside triphosphate. The physiologically significant donor in nearly all instances is probably ATP, although several protein kinases can also use GTP effectively in vitro. The last category includes caseine kinase 2, histone H1 kinase from tumor cells and p60[src]. Most of the produced protein-bound phosphate found in cells is present as phosphoserine and phosphothreonine and is formed as a result of the action of protein serine/threonine kinases. A much smaller fraction is present as phosphotyrosine and arises as a result of the action of protein tyrosine kinases. In addition to protein kinases that catalyze the previously mentioned phosphorylations, there are indications that protein kinases exist which are able to catalyze the formation of phosphohydroxylysine, phosphohistidine and phospholysine.[214]

Protein kinases can be initially divided into broad classes based on whether the amino acid acceptor of the phosphoryl residue is a protein alcohol group (i.e., the protein serine/threonine kinases, protein tyrosine kinases or hydroxylysine kinases) or a protein nitrogen-containing group (i.e., the protein histidine and lysine kinases). Within the main classes of kinases, individual groups of enzymes exist which can be characterized by their type of regulation. Protein serine and threonine kinases include groups regulated by cyclic nucleotides (e.g., cyclic AMP- and cyclic GMP-dependent protein kinases), Ca^{2+} and calmodulin-dependent protein kinases, diacylglycerol-dependent protein kinases or protein kinases C, double-stranded RNA-dependent protein kinases. Another level of classification is that both protein serine/threonine and tyrosine kinases can be divided into receptor- and non-receptor types.

CYCLIC AMP-DEPENDENT PROTEIN KINASES

Cyclic AMP-dependent protein kinases (protein kinases A) are enzymes activated by increased levels of intracellular cAMP. Such elevation in cAMP content is generally accomplished by hormonal stimulation of adenylyl cyclase.

Protein kinases A have been shown to represent a complex of two regulatory (R) subunits and two catalytic (C) subunits bound together in the absence of cAMP (Fig. 12). Two main isoforms of this holoenzyme were initially classified as type 1 and type 2 on the basis of their different R subunits.

The N-terminal one-third of the R subunit shows minor sequence homology between the different R subunits. The first 40-50 amino acid residues include the site responsible for dimerization. At approximately 90-100 amino acid residues from the N-terminus of the protein there is a more conserved, proteolitically sensitive hinge region which is the site of interaction with the C subunit. The type 1 R subunits have in this part of a molecule a pseudophosphorylation site R-R-G-A/G-V/I as well as a high-affinity binding site for MgATP. The type 2 R subunits contain a phosphorylation sequence in this hinge region R-R-X-S-V.[371] It was suggested that the R subunit inhibits the activity of the C subunit by binding of pseudophosphorylation or phosphorylation sites to the active site of the protein kinase. The C-terminal two-thirds of the type 1 and 2 R subunits display a high order of sequence

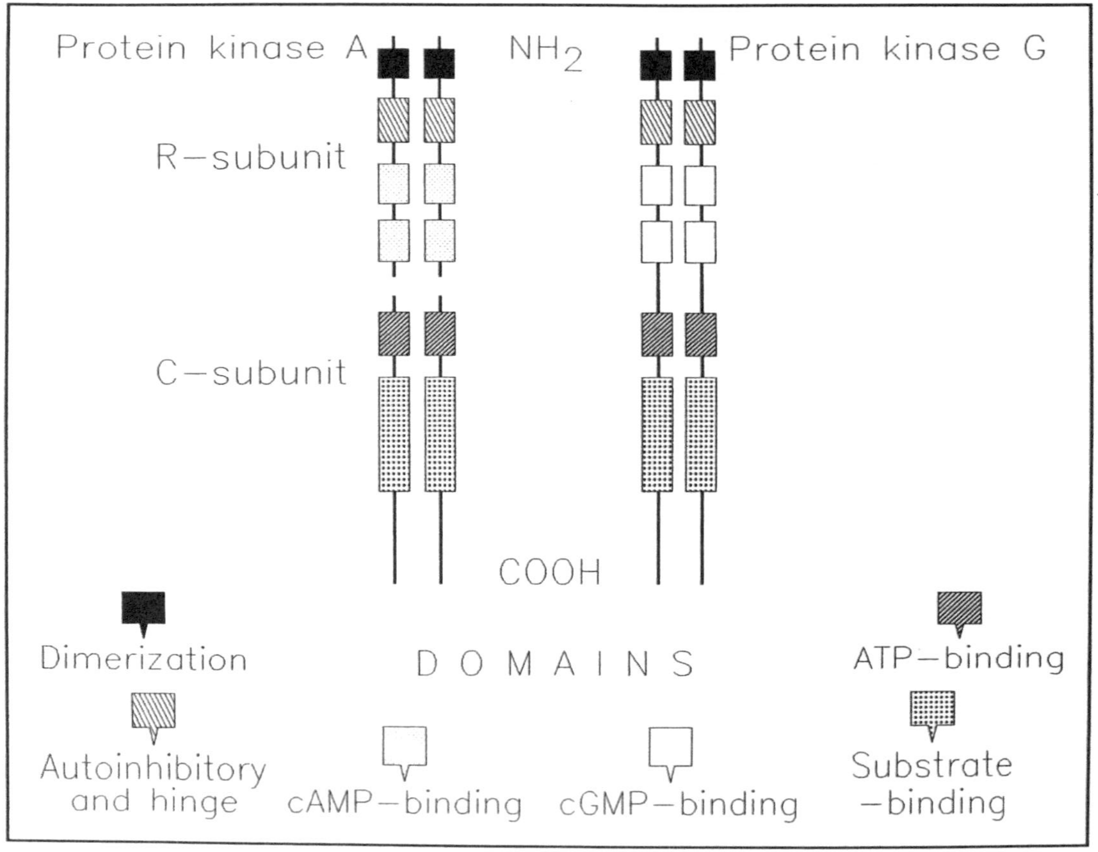

Fig. 12. Structural organization of cAMP- and cGMP-dependent protein kinases. cAMP-dependent protein kinase (protein kinase A) consists of two paired R and C subunits, whereas cGMP-dependent protein kinase (protein kinase G) represents a homodimer.

conservation. This segment contains two tandem cAMP-binding domains.

The catalytic domains of all protein kinases A are highly homologous. The N-terminal portion of the C subunit contains the ATP-binding site characterized by the motif G-X-G-X-X-G. This motif is conserved in many nucleotide-binding proteins and can be folded into an elbow around the nucleotide such that the G50 residue contacts the ribose moiety of ATP. Directly downstream from this sequence in all protein kinases is an invariant lysine residue K72 which appears to be directly involved in the phospho-transfer reaction.[194] The central portion of the C subunit is probably involved in both catalysis and substrate binding. An invariable D184 takes part in the phospho-transfer reaction, whereas the E170 residue and the surrounding sequence are probably involved in substrate interaction.

The process of protein kinase A activation is supposed to begin with the binding of cAMP to the R subunit. Following binding, the inactive holoenzyme dissociates into dimeric R subunits and two active C subunits. The latter are then free to catalyze the phosphorylation of a number of specific protein substrates containing the consensus sequences R-R-X-S/T-X. Protein kinase A activation is terminated by the hydrolysis of cAMP by specific phosphodiesterases. The released dimeric R subunits bind to C subunits and neutralize their activity.[213]

CYCLIC GMP-DEPENDENT PROTEIN KINASES

Unlike cAMP, cGMP interacts with several different intracellular receptor proteins including protein kinases, cyclic nucleotide phosphodiesterases, ion channels etc.[94,229] cGMP-dependent protein kinases (protein kinases G) are one of the major receptor proteins for cGMP in many tissues, in particular smooth muscle cells, where it is found in relatively high concentrations. The physiological substrates for cGMP kinases are however not well-defined in many cases, although several studies have demonstrated that cAMP and cGMP kinases catalyze the phosphorylation of the same proteins in vitro and in the intact cell.[88] Thus, the selective phosphorylation of these targets by protein kinases G requires the strict colocalization of cGMP kinase and the protein substrate. This strict compartmentalization could be important in determining what protein kinase in particular phosphorylates the particular substrate.

Most of the properties of protein kinases G have been investigated during studies on the purified bovine lung enzyme. It was shown that this enzyme has a dimeric structure which is due mainly to interactions of identical monomers, each of ca. 76 kDa, in their amino-terminal parts, called the dimerization domains (Fig. 12). The first 39 amino acid residues of the dimerization domain were observed to form a leucine/isoleucine repeat reminiscent of a leucine zipper motif, which has been shown to promote dimerization in a number of other proteins. Protein kinase G also contains, in addition to the region of dimerization of the holoenzyme, autophosphorylation sites, an autoinhibitory region, a hinge region and a region which determines cooperativity between the cGMP binding sites. The autoinhibitory domain, located between amino acids 54-67 for protein kinase G, inhibits protein kinase activity in the absence of cGMP.[228] Through studies of proteolytic fragments of this protein kinase it has been established that the inhibitory domain interacts primarily, or entirely, with the catalytic domain of the same protein chain rather than with the catalytic domain of the other chain of the dimer.[402]

Adjacent and C-terminal to the inhibitory domain, which ends at approximately 110 amino acid residues, are two tandem cGMP binding sites. Each of these positively cooperating sites contains approximately 120 amino acids and, although they exhibit different kinetics and cGMP analog specificity, are both involved in cGMP-dependent activation of protein kinase. cGMP binding to each site is supposed to interrupt the influence of the inhibitory domain on the catalytic domain. Another carboxy terminal segment of protein kinase G includes the Mg/ATP-binding site and the substrate binding site/catalytic domain. The catalytic domain is composed of approximately 330 amino acid residues and shows considerable homology with all other protein kinases described.

PROTEIN KINASES C

Protein kinase C (PKC) was originally detected as a protease-activated precursor for a cofactor-independent serine/threonine kinase known as protein kinase M.[368] Shortly thereafter, intact PKC was found also to exhibit kinase activity when it was associated with membranes.

It has been established that PKC is not a single molecular item and that many closely related protein kinase C isotypes exist, perhaps providing an explanation for the wide range of processes in which PKC has been implicated.[97,213] At present the mammalian PKC family consists of 12 different polypeptides: α, β1, β2, γ, δ, ε, ζ, η, θ, ι, λ and μ. In addition many nonmammalian PKCs have also been identified.

Sequence analysis demonstrates four regions of homology among the various isozymes: constant domains C1-C4, separated by variable regions V1-V5. The region C1 contains repeats of cysteine-rich sequence. This cysteine-rich sequence is essential for the binding of phorbol ester and possibly also diacylglycerol.[285] The N-terminal part of the C1 domain contains a basic-rich sequence similar to the phosphorylation sequence for many substrates except that an alanine residue replaces a phosphorylatable serine or threonine: R-X-X-A-X-R.[149] It was proposed that this region is a pseudosubstrate domain and serves an autoinhibitory function. Thus synthetic peptides with this sequence were shown to inhibit protein kinase C, whereas peptides in which the alanine is replaced with serine or threonine residues serve as substrates.[175] It was also hypothesized that cofactors or substrates may activate the enzyme by inducing or stabilizing an "open" enzyme conformation in which the inhibitory domain has been removed from the active site. The C2 domain appears to be involved in Ca^{2+} regulation since isozymes lacking the C2 domain have been found to be Ca^{2+}-independent. The C3 domain contains a consensus ATP binding sequence while the C4 domain is hypothesized to be involved in substrate recognition.

The hydrolysis of protein kinase C by calpain or trypsin in the V3 region generates the C-terminal fragment (known as protein kinase M) containing C3 and C4 domains, and an N-terminal regulatory fragment with C1 and C2 domains. The C-terminal fragment retains the proteolytic activity whereas the N-terminal domain preserves membrane binding functions.[328] It is probable that calpain-dependent proteolysis has some importance under in vivo conditions and initiates the degradation of the activated form of PKC molecules. This proposition is based on an observation that such a proteinase is activated at the micromolar range of Ca^{2+} and effectively cleaves PKC in the presence of phosphatidylserine and DAG.

The main regulator of protein kinase C activity is confirmed to be diacylglycerol. It is generally produced by activation of type C phospholipases, which hydrolyze lipids, particularly phosphatidylinositol-4,5-diphosphate, to generate DAG and inositol-1,4,5-triphosphate. Another important source of DAG is phosphatidylcholine, which can be hydrolyzed to diacylglycerol and phosphocholine. In the presence of phospholipids DAG leads to activation of PKC. In the case of the Ca^{2+}-dependent forms of PKC this effect is seen to be further augmented by IP_3, which promotes a rise in intracellular Ca^{2+} levels.

Evidence from many laboratories demonstrates the importance of the free hydroxyl and both carbonyl groups of DAG for protein kinase C activation. In addition, the hydrophobic domain of diacylglycerols, which is provided primarily by a long acyl chain in the 1-position and an equal or shorter one in the 2-position is also necessary for regulatory activity.

It was observed that DAG requirements for activation of PKC could be replaced by phorbol ester tumor promoters. This provided the suggestion that protein kinase C might be the phorbol ester receptor. Several lines of evidence supported this hypothesis. First, phorbol esters regulated the translocation of PKC from cytosol to membrane. Second, both phorbol ester binding and PKC activation were shown to depend on similar cofactor requirements. Third, diacylglycerol competed for phorbol ester binding.[328] Other tumor-promoting compounds of diverse structure have also been found to activate PKC. These include teleocidins, aplysiatoxins, bryostatins, mezereins.

Not all protein kinases C are, however, regulated by diacylglycerol. For example, PKC-ζ (and possibly PKC-λ) are not activated by phorbol esters/DAG. There is evidence that these isoforms of PKCs are stimulated by phosphatidylinositol (3,4,5)-triphosphate. Such behavior is of great interest since it indicates that the signal transduction pathways in which PKC isotypes may be involved in vivo are not restricted solely to those leading to the generation of

DAG but might also include those that activate the PI 3-kinase pathway.

The calcium requirement for PKC activation was demonstrated for isozymes α, β and γ. Protein kinase C δ, ε, ζ, θ and η lack the Ca^{2+} binding C-2 domain and therefore are not dependent on Ca^{2+} ions. The mechanism of Ca^{2+} stimulation is probably linked to increased tightness of association of protein kinases C with the membrane, thereby increasing the membrane occupancy and binding of DAG. Normally PKC is folded so that an endogenous "pseudosubstrate" region of the protein is bound to the catalytic site, thereby inhibiting its activity. The combination of DAG and Ca^{2+} causes a conformational change in PKC, causing flexing at a hinge region so as to withdraw the pseudosubstrate and unblock the catalytic site. The colocalization of DAG and protein kinase C in membranes is usually transient and therefore protein kinase C is activated for only a short time after a receptor has been stimulated.[410]

Early kinetic and phorbol ester binding studies indicated that protein kinase C required acidic phospholipids for full activity. Phosphatidylserine, phosphatidic acid and phosphatidylinositol were the most potent, whereas phosphatidylcholine and phosphatidylethanolamine were ineffective. A specific site on PKC for interaction with phospholipids has not yet been identified. However the possible domains which could provide binding function may include the regulatory fragment and the catalytic region.

CA²⁺/CALMODULIN-DEPENDENT PROTEIN KINASES

Several types of Ca^{2+}/calmodulin-dependent protein kinases (CaM kinases) have been characterized in mammalian systems. These are phosphorylase kinase, myosin light chain kinase and CaM kinases 1, 2 and 3.[276] They have native molecular weights of 250-650 kDa and are composed of α and β subunits of 50-60 kDa.

CaM kinase 2 is the typical representative of these enzymes. Among its substrates are synapsin I, microtubule-associated protein-2 (MAP-2), glycogen synthase, tyrosine hydroxylase and smooth muscle myosin light chain. Comparison of the known amino acid sequences of the phosphorylation sites in these proteins, as well as studies using synthetic peptides, have revealed a consensus sequence of R-X-X-S/T as a minimum requirement for phosphorylation by CaM kinase 2.

One of the interesting features of CaM kinase is its predisposition to autophosphorylation. As a result, such in vitro autophosphorylation converts the enzyme to a species which is no longer regulated by Ca^{2+}/calmodulin. This effect can be reversed by treatment with phosphatase. The time course of the conversion of the enzyme to a Ca^{2+}/calmodulin-independent form correlates with autophosphorylation by an intramolecular mechanism involving threonyl residues within both the α and β subunits of the protein kinase. Detailed investigations have identified threonine 286 in the α subunit or threonine 287 in the β subunit as the autophosphorylation residues. The surrounding sequence, R-Q-E-T, conforms to the consensus phosphorylation sequence of the enzyme. Since the modifiable threonine is adjacent to the predicted CaM-binding domain in each subunit of the enzyme, one can suggest that autophosphorylation may maintain a structure which is functionally equivalent to that which occurs following the binding of Ca^{2+}/calmodulin.

Almost complete conversion of the enzyme to the Ca^{2+}/calmodulin-independent form can be achieved with the phosphorylation of only 3-4 of the possible 10-12 subunits. These data suggest cooperative interaction between phosphorylated and unphosphorylated subunits. The biological importance of the generation of a Ca^{2+}/calmodulin-independent form of CaM kinase 2 includes a mechanism which could allow the enzyme to remain active in cells even after Ca^{2+} levels have returned to basal levels. In such a case, some phosphatase could represent a signaling component necessary for terminating the protein kinase activity.[276]

THE SRC FAMILY OF TYROSINE PROTEIN KINASES

Among the proteins involved in signal transduction, the tyrosine protein kinases (TPKs) appear to play key roles in the initiation and continuation of various signaling cascades. The known TPKs can be divided into two groups based on their predicted structure. The kinases of the largest group contain extracellular and transmembrane domains and in many cases are known to bind polypeptide hormones (receptor TPKs). The second group lacks both sequences and is considered to be non-receptor TPKs, although many of these enzymes are known to associate with various cell surface receptors. This TPK group consists of at least eight different families[385] and includes Src as a typical representative.

At least nine family members of Src PTKs have been identified. Four of them, Src, Yes, Fyn and Lyn, are expressed in a variety of cells. In contrast, Lck, Hck, Fgr, and Blk are primarily expressed in hemopoietic cells. The Yrk kinase is expressed at the highest levels in the brain and spleen.

All Src PTKs have very similar structures (Fig. 13). The N-terminus of a typical Src family member contains the sequences necessary for myristoylation and reversible palmitoylation.[291] These modifications are required for the stable membrane association of Src TPK and are necessary for proper regulation of enzymatic activity.[12,43] Adjacent to this small fragment is the 50- to 80-residue-long "unique" domain. In this part the amino acid sequences of different Src TPKs differ most dramatically from each other. The next region located approximately between amino acid residues 90 and 250 represents a sequence sharing significant homology with different Src TPKs as well as numerous other proteins. This region can be subdivided into two domains: Src homology 3 (SH3) and Src homology 2 (SH2) which have lengths of about 50 and 100 amino acid residues respectively. Both SH2 and SH3 are important for regulating the signal-driven translocations and enzymatic activity of Src TPKs. The contribution of SH2 is due to its ability to specifically bind phosphotyrosine-containing peptides. Although the structure of SH3 and the SH3-binding sequences have been identified, the exact mechanism which mediates its effect upon the Src TPK activity remains unclear in many details. One confirmed proposition is that SH3 affects the activity of Src TPKs by mediating their binding to cytoskeletal as well as other cellular proteins containing proline-rich sequences.

The major part of the C-terminal half of the Src TPKs represents the catalytic domain (SH1), which is the region of the highest homology between the Src family members. This domain contains the sites for ATP binding and tyrosine autophosphorylation and is thought to be critical for the activation of the Src TPKs. The last 17 to 19 amino acid residues represent the Src regulatory

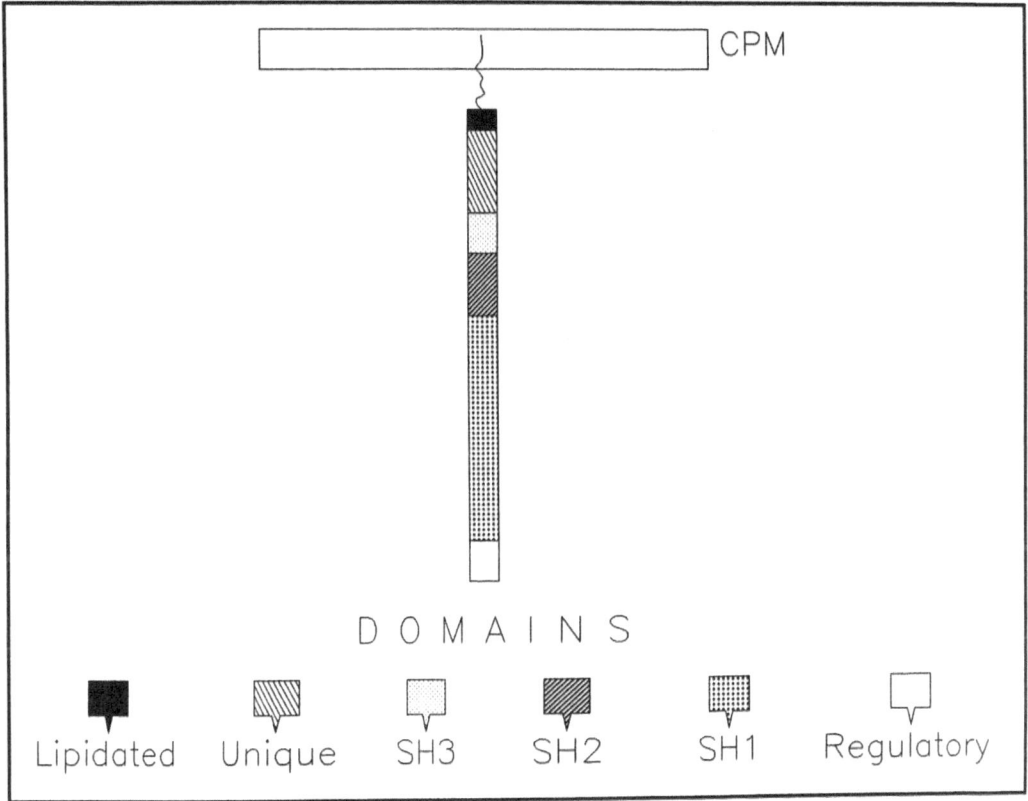

Fig. 13. Structural organization of protein tyrosine kinases of the Src family. CPM, cytoplasmic membrane. Near the N-terminus of Src protein kinases there are sites for myristoylation and palmitoylation, which are necessary for attachment of the enzyme molecules to CPM.

domain and include a tyrosine residue. This is the major site of phosphorylation in the Src TPKs with a generally negative effect on protein kinase activity.

The regulation of Src TPK enzymatic activity is thought to be primarily mediated by phosphorylation. The Src kinases are believed to transphosphorylate each other on tyrosine residues within the catalytic region. This kind of autophosphorylation is thought to contribute to an increase in the activity of Src TPKs. Since Src kinases in many instances are associated with specific receptors, it is likely that the ligand-induced aggregation of receptors contributes to the co-apposition of the enzymes. In contrast, the C-terminal negative regulatory site is phosphorylated, usually by another class of cellular TPKs.[373] The proposed mechanism of such inhibition is that a phosphotyrosine residue interacts with the SH2 domain of the same molecule. Such binding is thought to change the structure of Src-related TPK and arrest its interactions with relevant protein substrates.

Both sites of tyrosine phosphorylation are susceptible to the action of tyrosine protein phosphatases which can reverse the effects of tyrosine phosphorylation on the function of Src TPKs. CD45, a receptor-like protein tyrosine phosphatase, is an example of a phosphatase which interacts with Src and activates this enzyme by dephosphorylating the C-terminal tyrosine residue.[248] Thus Src TPK activity is balanced between the action of cellular kinases and phosphatases.

SH2 AND SH3 DOMAINS

Eukaryotic cellular signal-transduction pathways that are linked to tyrosine kinases rely in many instances on two small protein modules, known as SH2 and SH3 domains (Src homology domains). The key aspect of the function of SH2 and SH3 domains is their ability to recognize particular amino acid sequences in their target proteins: SH2 domains bind tightly to phosphorylated tyrosine residues whereas SH3 domains bind peptide sequences that are rich in proline and hydrophobic amino acids. Although SH2 and SH3 domains frequently occur close together in a sequence, some proteins have only one or the other domain and some have more than one version of either domain. The N- and C-terminal ends of the SH2 and SH3 domains are closed together in globular structures, which allows them to protrude from the rest of the protein and function independently.

Proteins that contain SH2 and SH3 domains fall into two broad classes: those with catalytic functions and those without. Enzymes that contain these domains include cytoplasmic tyrosine kinases, phosphotyrosine phosphatases, phospholipases Cγ, Ras GAPs, Ras GDIs, etc.[208] The SH2 and SH3 domains in these proteins serve to modulate catalytic activity, or/and target the enzymes to certain cellular locations. The second class of SH2 and SH3-containing proteins do not exhibit enzymatic activity. These proteins are comprised almost exclusively of SH2 and SH3 domains (Crk, Nck, Sem-5/Grb2) and are known as "adapter" proteins.[295] Thus adapter proteins represent some sort of "bridge" between signaling components.

The sequence specificity of phosphotyrosine-SH2 interactions was first determined for the p85 subunit of phosphatidylinositol-3 kinase, Ras GAP and PLCγ. These studies showed that relatively short peptide sequences are sufficient to bind the SH2 domain with high affinity. For example p85 binds preferentially to peptides with methionine or valine at the first position followed by the phosphotyrosine and methionine at the third position. In contrast, GAP binds to a site on the PDGF receptor with methionine and proline at the first and third positions respectively. In many instances three residues immediately C-terminal to the phosphotyrosine were shown to be the most important in determining the binding affinities of phosphopeptides. For example, SH2 domains of Src family tyrosine kinases are attracted by the sequence Y*EEI (more generally, phosphotyrosine followed by two non-basic polar residues and a large hydrophobic residue).

Recent structural data indicate that SH2 domains have a conserved binding pocket for phosphotyrosine and a pocket for the amino acid three residues C-terminal to the phosphotyrosine (the +3 position). This latter pocket is lined by variable residues and allows different SH2 domains to bind preferentially to distinct residues at the +3 site. SH2 domains also have specific contact sites on their surface for the residues at the +1 and +2 positions. Thus phosphorylation of the tyrosine residue is thought to induce SH2 association, while the +1, +2 and +3 C-terminal residues provide for specific recognition of the relevant SH2 domain.

THE MAP KINASE CASCADE

The mitogen-activated protein kinase (MAPK) pathway is a conserved eukaryotic signaling module that converts receptor signals into a variety of outputs via trans-cytoplasmic signaling to the nucleus. In the nucleus, transcription of specific genes is induced through phosphorylation and activation of transcription factors. Currently four distinct MAPK pathways are known in budding yeast and three in vertebrates, but others may yet be found.[181]

Mitogen-activated protein kinases are central components of the MAPK pathway and represent a group of 40-65 kDa serine/threonine kinases. Several different members of the MAPK pathway have been cloned from various animal species, including rat, mouse, man, *Xenopus laevis* and starfish, and characterized.[190] MAP kinases become enzymatically activated by phosphorylation on both threonine and tyrosine residues in response to various extracellular stimuli. The sites of threonine and tyrosine phosphorylation have been identified and are located one residue apart in the consensus kinase sequence on a loop that is involved in substrate binding.[5] MAP kinase could be dephosphorylated on phosphothreonine by treatment with the serine/threonine-specific protein phosphatase 2A, whereas treatment with the tyrosine-specific phosphatase CD45 leads to specific dephosphorylation of phosphotyrosine residues. In both cases complete inactivation of MAP kinase activity was observed.[8]

One of the well defined MAPK pathways is that beginning with activation of growth factor receptor by growth hormone (Fig. 14). Attachment of a ligand to this receptor protein tyrosine kinase induces translocation of the complex of Grb2 and Ras GDS (the latter is also known as SOS) to the plasma membrane. This process is driven by binding of GRB2 to an autophosphorylation site in the RPTK itself or on some docking protein phosphorylated by protein kinase.[387] Juxtaposition of SOS and Ras low molecular weight GTP-binding protein at the plasma membrane results in exchange of Ras-GDP for Ras-GTP and leads to Ras activation. GTP-bound Ras binds to the N-terminus of Raf protein

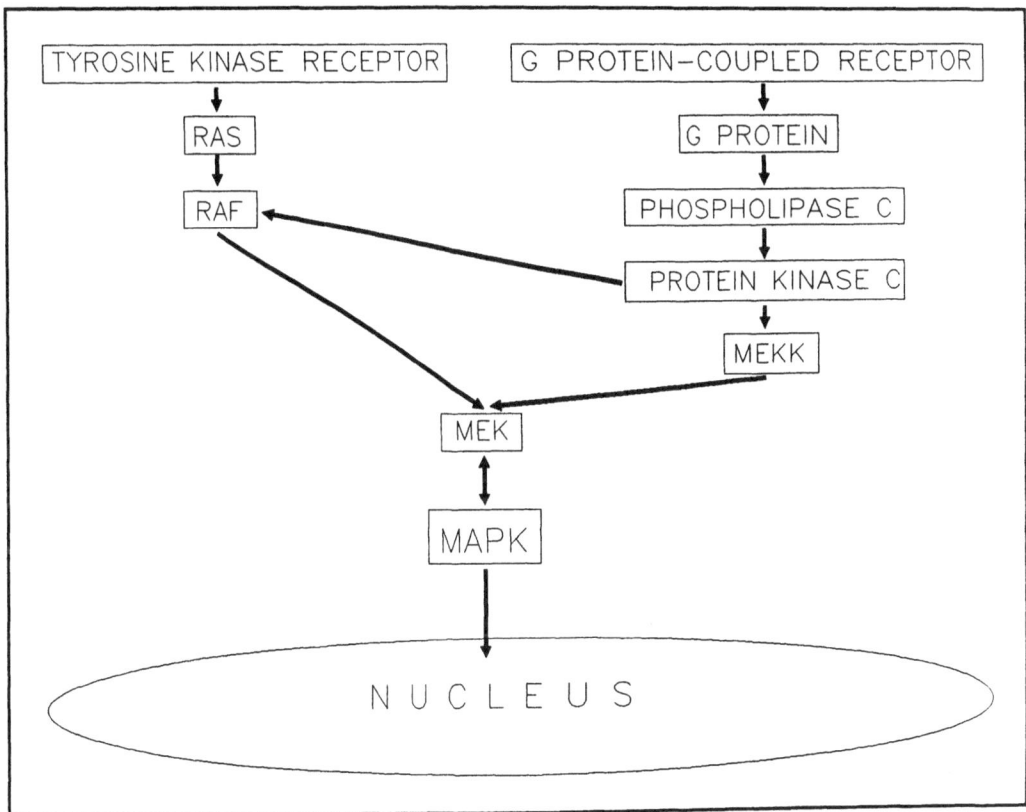

Fig. 14. MAP kinase cascades. MAPK, mitogen-activated protein kinase; MEK, MAPK kinase; MEKK, MAPK kinase kinase. For details see text.

serine/threonine kinase family members, thus bringing Raf to the membrane.

The next step in signal transduction is phosphorylation of a 45 kDa dual-specificity tyrosine/threonine protein kinase known as MAP kinase kinase or MEK. Like MAP kinase, MEK is inactivated upon treatment with protein phosphatases and complete inactivation occurs with the serine/threonine phosphatases 1 and 2A. In spite of the fact that MEK autophosphorylates on tyrosine residues in vitro, tyrosine phosphorylation was not found in situ and tyrosine phosphatases had no effect on enzyme activity. MEK has a remarkably narrow substrate specificity in vitro, restricted so far solely to MAPK. In contrast, MEK itself could be phosphorylated by different protein kinases and it serves as a convergence point at which multiple signaling pathways meet.[298]

The phosphorylation of MAPK by MEK activates the former protein kinase and triggers the phosphorylation of numerous cellular proteins. Substrates of MAP kinase are various and many of them have been found to be important regulatory proteins. Several transcription factors, such as *c-jun, c-myc,* p62[TCF] are phosphorylated in vitro by MAPK. A physiologic substrate of MAPK is the serine/threonine kinase pp90[rsk], also known as S6 kinase 2. S6 kinase 2 is implicated in the phosphorylation of ribosomal protein S6 during both meiotic maturation and the mitogenic response. Another target of MAPK in vivo may be MAPK itself.[99,213]

Recently cytoplasmic PLA$_2$ was identified as another substrate of MAPK.[227] Stimulated cPLA$_2$ is translocated from the cytoplasm to membranes where it cleaves phospholipids and produces potent second messengers that mediate numerous biological responses. It has been proposed that the rapid activation of cPLA$_2$, induced by agents such as phorbol esters, is achieved by the synergistic action of Ca^{2+} and phosphorylation of cPLA$_2$ by MAP kinase. Among other target proteins for MAP kinase phosphorylation appears to be the EGF receptor, which seems to be both an activator and, through a feed-back mechanism, also a substrate of MAPK.[190]

Mitogen-induced activation of the MAPK pathway may also be achieved by several other routes, involving the activation of various isoforms of PLC leading to activation of protein kinase C α, β, and γ. These PKCs are supposed to phosphorylate and activate Raf directly,[212] completing the evidence for a pathway leading from G protein and phospholipase C, via protein kinase C to Raf, MEK and MAPK. MEK can also be activated independently of Raf by a MEK kinase (MEKK), which might represent another substrate of protein kinase C.[218] It can be concluded therefore that Raf might mediate extracellular signals in general from receptors with tyrosine kinase activity, whereas MEKK may regulate the MAP kinase network via receptors that activate G proteins and protein kinase C.

Taken together, these studies indicate that MAP kinases may serve as a central point in a signal transduction network where different upstream pathways converge and may distribute the signals to different downstream targets.

CHAPTER 9

PROTEIN
PHOSPHATASES

As mentioned in previous sections, phosphate ester formation and hydrolysis are likely the most prevalent chemical reactions carried out by living organisms. These reactions regulate numerous biological processes and are catalyzed by protein kinases and protein phosphatases respectively. In eukaryotic organisms phosphatases are typically categorized based upon substrate specificity. Three main groups of phosphatases exist: the nonspecific phosphatases, serine/threonine phosphatases and tyrosine phosphatases.[359]

The nonspecific phosphatases include enzymes such as the alkaline phosphatases and the acid phosphatases. These enzymes are often capable of hydrolyzing both proteinaceous and nonproteinaceous phosphate monoesters and are supposed to not participate in signal transduction directly. In turn, the nonspecific phosphatases typically function in degradative or catabolic pathways and are believed to serve the function of "phosphate scavengers".

Members of protein serine/threonine phosphatases (PSPs) are specific for dephosphorylation of serine/threonine residues (however at least one protein serine/threonine phosphatase type 2A has a low level of tyrosine phosphatase activity as well).[73] The number of distinct forms of PSPs that have been so far identified (ca. 30 members) is considerably smaller than the number of protein serine/threonine kinases. While the numbers of characterized phosphatase genes will undoubtedly increase, one should conclude that a single phosphatase catalyzes the dephosphorylation of proteins phosphorylated by more than one protein kinase.

The classification of the protein serine/threonine phosphatases is a confusing issue and there is not yet a general consensus on the best system. The first classification scheme and the one that is used most widely is based on biochemical differences in the activity and regulation of individual enzymes. The criteria for grouping individual enzymes into classes include the relative activity toward the α and β subunits of phosphorylase kinase and sensitivity to two of the inhibitor proteins, termed inhibitor 1 and inhibitor 2. Type 1 protein phosphatases (PSP1) preferentially dephosphorylate the β subunit of phosphorylase kinase and are sensitive to both inhibitor 1 and 2. The type 2 enzymes preferentially dephosphorylate the α subunit of phosphorylase kinase and are insensitive to either of the inhibitor proteins. The type 2 enzymes are further divided on the basis of structural and regulatory properties. Type 2A phosphatase (PSP2A) represents a class of oligomeric enzymes with no obvious requirements for ions or cofactors. Type 2B phosphatases (PSP2B) are regulated by Ca^{2+}/calmodulin. Another class of protein serine/threonine phosphatases, type 2C, requires Mg^{2+} for activity. This functional classification has been confirmed in general during cloning of individual forms of protein serine/threonine phosphatases and comparison of degree of relatedness in corresponding sequences.[210]

In addition to protein serine/threonine phosphatases identified by biochemical methods, a number of proteins related to the serine/threonine phosphatase family have been identified using molecular genetic methods. Although they are clearly members of the family, the predicted sequences of these proteins are sufficiently different and they cannot be easily classified into one of the classes described above. Because many of these forms have only been identified through DNA cloning, the biochemical properties of the individual forms have not yet been characterized.

The protein tyrosine phosphatases (PTPs) include a collection of over 40 enzymes. They share no sequence similarity with the PSPs or nonspecific phosphatases. The PTPs are often categorized into three groups: (1) receptor-like PTPs; (2) intracellular PTPs; and (3) dual specificity PTPs.[359] The latter group is unique in its ability to utilize phosphoserine and phosphothreonine as substrates in addition to phosphotyrosine.[71,380]

PROTEIN SERINE/THREONINE PHOSPHATASES

PROTEIN PHOSPHATASES 1

Type 1 protein phosphatases (PSP1) are multimeric structures composed of a catalytic subunit complexed to a number of accessory subunits, such as inhibitors or a glycogen binding protein termed the G subunit.[360] The free catalytic subunit has not been detected in cell or tissue extracts, suggesting that little if any uncomplexed catalytic subunits are present in vivo.

The activity of PSP1 is regulated in vitro by interaction of the catalytic subunit with endogenous regulatory subunits and inhibitory proteins. Three heat-stable inhibitor proteins, inhibitor 1, DARPP-32 and inhibitor 2, have been identified in many tissues. The activity of the PSP1 catalytic subunit is also controlled by interaction with the G subunit which serves to stimulate phosphatase activity and allows its association with the protein-glycogen complex in skeletal muscle.

Inactivation of PSP1 by inhibitor 1 is dependent on phosphorylation of the inhibitor by cAMP-dependent protein kinase.[6] Therefore this protein provides a potential mechanism through which increases in intracellular cAMP can decrease the activity of PSP1. Inhibitor 1 is poorly dephosphorylated by PSP1, especially in the presence of other phosphoproteins and physiological concentrations of cations.[279] In contrast, it is readily dephosphorylated by PSP2A and PSP2B, suggesting that different protein phosphatases may be capable of cross-talk through dephosphorylation of this inhibitor. In addition, activation of PSP2B by Ca^{2+} and subsequent dephosphorylation of inhibitor 1 could provide a mechanism for antagonism of cAMP-dependent inhibition of PSP1.

The protein DARPP-32 is another inhibitor related to inhibitor 1 that is enriched in dopamine-innervated regions of the brain. Like inhibitor 1, DARPP-32 is phosphorylated in response to dopamine or other agents that elevate cAMP and is only active after phosphorylation on a critical threonine residue.[165] The N-terminal region of DARPP-32, which includes the phosphorylated threonine residue (T340), contains the inhibitory domain. The inhibitory threonine residue is specifically dephosphorylated by PSP2B, which is also localized in the same brain regions as DARPP-32.

Inhibitor 2 is a third type of PSP1 inhibitor protein that is also referred to as a modulator protein. Despite certain similarities in physical properties (e.g., heat- and acid-stabilities) there is no sequence homology between inhibitor 2 and inhibitor 1 or DARPP-32. Inhibitor 2 does not require phosphorylation for PSP1 inhibitory activity and it has two distinct effects on the PSP1 catalytic subunit. The first one is rapid and accomplishes inhibition of activity by competition for peptide substrate binding. The second is a time-dependent conformational change in the catalytic subunit of PSP1 resulting in enzyme inactivation. The inactive PSP1/inhibitor 2 complex can then be reactivated by phosphorylation of inhibitor 2 by glycogen synthase kinase. Phosphorylation of T72 of inhibitor 2 by this protein kinase is thought to initiate a reactivating process resulting in a conformational change.[46] Another enzyme which participates in modification of inhibitor 2 is casein kinase 2. Inhibitor 2 is phosphorylated in vitro by the latter enzyme at several sites.[168] These phosphorylations cause an increase in phosphorylation of T72 by glycogen synthase kinase and a stimulation of the reactivating reaction. The level of inhibitor 2 has been shown to fluctuate during cell growth. For example, the level of inhibitor 2 protein and inhibitor 2 activity in rat fibroblasts oscillates with each cell cycle, peaking during both S phase and mitosis.[49]

A fourth set of PSP1 inhibitors has recently been identified in the nuclei of bovine thymus cells.[31] Two 18 kDa and 16 kDa inhibitors are active against PSP1 but not PSP2A, PSP2B or PSP2C. These products block the activity of PSP1 specifically, i.e., toward phosphorylase kinase and casein but not toward myelin basic protein.

PROTEIN PHOSPHATASES 2A

Type 2A protein phosphatase has been isolated from mammalian tissues as multi-subunit complexes. These complexes are composed of different combinations of three subunits termed A, B and C, where C is a catalytic component. PSP2A is thought to consist in vivo primarily of the ABC heterotrimeric form. Dissociation-reconstitution studies have demonstrated that both the A and B subunits play roles in controlling the activity and specificity of the catalytic subunit. Recent evidence indicates that while the A and C subunits are expressed constitutively, the B subunits are expressed differentially. This differential expression of different classes and isoforms of the B subunits is likely to be one of major mechanisms for long-term regulation of PSP2A. In addition there is also evidence for potential short-term regulation of PSP2A activity by posttranslational modification reactions.

The free catalytic subunit or the heterodimeric AC form of PSP2A can be phosphorylated in vitro by protein kinases including Src, Lck, the epidermal growth factor receptor and the insulin receptor.[72] Phosphorylation of the PSP2A catalytic subunit is enhanced by the inclusion of okadaic acid, suggesting that the protein is rapidly autodephosphorylated. Phosphorylation occurs on Y307 at the C-terminus of the protein and thiophosphorylation at this site causes a 90% inhibition of phosphatase activity. Due to these phosphorylation-dephosphorylation reactions phosphatase 2A could be connected to the signaling system of eukaryotic cells.

PROTEIN PHOSPHATASES 2B

Type 2B protein phosphatase (PSP2B) was initially identified as a calmodulin-binding protein that could inhibit the activity of calmodulin-stimulated cyclic nucleotide phosphodiesterase.[206] The protein was also found to bind Ca^{2+} and was called calcineurin due to a predominant localization in mammalian nervous tissue. The native form of PSP2B consists of a 1:1 molar complex between a 60-61 kDa catalytic (A) subunit and a 19 kDa regulatory (B) subunit.

The predicted primary structure of the PSP2B catalytic subunits contains a protein serine/threonine phosphatase catalytic domain that is 40-50% identical to those of PSP1 and PSP2A. The catalytic domain of PSP2B is followed by a C-terminal extension of 180-200 amino acids that is absent in PSP1 and PSP2A. This region contains a consensus calmodulin binding site and putative sites for phosphorylation by protein kinase C and calmodulin-dependent protein kinase 2, which may be involved in regulation of phosphatase activity.[160] However, phosphorylation of PSP2B in vivo has not been demonstrated and its importance as a physiological regulatory mechanism is not clear.

The primary structure of the B subunit is similar to calmodulin and contains four putative Ca^{2+} binding sites. Following Ca^{2+} binding, the B subunit undergoes a conformational change that results in stimulation of the A subunit phosphatase activity. The activity of PSP2B is also regulated by calmodulin. In the presence of physiological concentrations of Ca^{2+} calmodulin forms a 1:1 molar complex with PSP2B and stimulates protein phosphatase activity.[205] Although the direct binding of Ca^{2+} to the B subunit may regulate phosphatase activity, under optimal assay conditions direct Ca^{2+} stimulation of the enzyme is relatively low. In contrast, interaction of the Ca^{2+}/calmodulin complex with PSP2B results in significant (10-fold) activation, suggesting that PSP2B is primarily a calmodulin-regulated protein phosphatase. Another mechanism which could modulate activity and/or subcellular localization of PSP2B is N-terminal myristoylation of the B subunit. The role of B subunit myristoylation has been confirmed for association of PSP2B with membranes in neural tissues.

PROTEIN PHOSPHATASES 2C

On the basis of activity toward commonly used phosphoprotein substrates, protein phosphatase 2C (PSP2C) is present in many tissues at much lower levels than the other types of protein serine/threonine phosphatase. Unlike the other protein serine/threonine phosphatases, purified PSP2C is a monomeric 43 kDa protein that has the unique property of requiring divalent metal ions for activity. The half-maximal concentration of Mg^{2+} required for PSP2C activation in tissue extracts is 1mM, which is similar to normal intracellular concentrations of this ion.

The predicted amino acid sequences of the PSP2C isoforms are very distinct from those of other members of the protein serine/threonine phosphatase family and do not contain an obvious phosphatase catalytic domain.[369] This indicates that PSP2C is a separate gene family and could have a catalytic mechanism distinct from that of the members of the PSP1/PSP2A/PSP2B gene family.

PROTEIN TYROSINE PHOSPHATASES

The first isolated protein tyrosine phosphatase was PTP1B. The structural analysis of this and related enzymes disclosed several functional domains. Although PTP1B was originally purified as a monomeric catalytic subunit of 37 kDa, isolation of cDNA for PTP1B predicted a longer protein of ca. 50 kDa with an extension of 114 amino acids at the C-terminus which serves a regulatory function. The extreme C-terminal 35 amino acid residues are both necessary and sufficient for targeting to membranes of the endoplasmic reticulum. The preceding 88 amino acids are predominantly hydrophilic and bear several putative sites for phosphorylation by serine/threonine kinases. The catalytic domain was shown to be located at the N-terminus of the molecule and accounts for approximately 55% of the protein sequence.[396]

The identity of physiological substrates for PTP1B is unknown. Its localization raises the possibility of a role in controlling the dynamic changes in endoplasmic reticulum structure and regulation of the cytoskeleton rearrangements during certain phases of the cell cycle. The demonstration of endoplasmic reticulum localization of tyrosine kinase Ltk also suggests that this tyrosine kinase or its substrate may be susceptible to the action of PTP1B in vivo.

PTP1C and SHP/HCP protein tyrosine phosphatases have been isolated from human and mouse tissues. Each protein contains a phosphatase domain in the C-terminal half of the protein and two SH2 domains at the amino-terminus. Since the SH2 domain in PTP1C can bind tyrosine-phosphorylated proteins in vitro, it probably allows the phosphatase to associate with particular proteins and then dephosphorylate them or other closely associated proteins.

Another motif identified in several PTPs has homology to cytoskeletal-associated proteins. This region of homology is located at the N-terminus of the proteins followed by a spacer of approximately 300 amino acid residues. Since many of the cytoskeletal proteins have been shown to be phosphorylated on tyrosine, they clearly represent possible targets for the cytoskeleton-associated protein tyrosine phosphatases. Although the precise function of these phosphorylation sites on cytoskeletal proteins has not been determined, changes in the phosphorylation state of these proteins have been shown to correlate with changes in the structure of the cytoskeleton.[396]

Two SH2 domain-containing PTPs have been studied in some detail, SH-PTP1 and SH-PTP2. Both proteins contain dual amino terminal SH2 domains and have been shown to interact with growth factor receptors and phosphorylated insulin receptor substrate via their SH2 domains. The intracellular location of SH-PTP1 changes in response to extracellular stimulus. Within 1 min of platelet stimulation by thrombin, SH-PTP1 becomes associated with the cytoskeleton. This mimics the translocation of tyrosine kinase Src in response to the same stimulus and implicates the participation of SH-PTP1 in the complex series of cytoskeletal rearrangements that accompany platelet activation. SH-PTP2 may also be involved in tyrosine kinase receptor transduction since this phosphatase can be bound by the Ras pathway adapter molecule Grb2 and can bind to and be phosphorylated by activated platelet-derived growth factor receptor. In addition to participation in translocations of SH-phosphatases, SH2 domains are known to regulate specific activity. For instance, removal of the SH2 domains from SH-PTP2 resulted in a 12- to 45-fold increase in phosphatase activity. Exogenous SH2 domains reverse this effect.[359]

CHAPTER 10

GENERAL COMMENTS ON SIGNAL TRANSDUCTION IN EUKARYOTIC CELLS

The data presented in the above part of the book summarize general knowledge on the receipt and processing of information necessary for an adequate response of eukaryotic cells to steadily changing environment. Computation by cells of information which is deposited in particular ligands starts with a binding of such primary messengers to, and activation of, certain receptors. These processes typically result in modification of activities of certain proteins (protein kinases and phosphatases) accomplishing direct effects on downstream targets, or in activation of enzymes and ion channels which produce or pump up accordingly secondary messengers necessary for transducing the signal further. The signal usually proceeds through several cascade pathways and ultimately reaches and alters activities of components which directly participate in generation of a specific response of the eukaryotic cell.

There are several main principles which form the basis of the machinery accomplishing signaling functions.

1. The first characteristic which attracts attention is the apparent functional complexity of signaling mechanisms. The accumulated literature data suggest that the process of intracellular signaling represents not a single vector connecting a cell surface receptor with a protein accomplishing a specific response, but a stream of

information along a dominant pathway. It contains points for divergence necessary for switching this flow onto and activation of other pathways, as well as convergence points to collect and sort the information from different routes. The fine tuning of this apparatus is accomplished by numerous feedback reactions and modulatory mechanisms which decrease or increase functional activities of key enzymes.

2. Another prominent feature of signaling machinery is the biochemical simplicity of intracellular signaling. This supposition will become obvious if one considers the fact that very few types of main secondary messengers (diacylglycerol, inositol 1,4,5-triphosphate, cyclic AMP, cyclic GMP, nitric oxide and Ca^{2+}) are produced intracellularly, compared to an extremely diverse spectrum of extracellular ligands (primary messengers).

3. In order to generate specific response to specific signal, eukaryotic cells possess mechanisms for compartmentalization of signal spreading. This means that generated second messengers or activated enzymes transducing the stimulus downstream accomplish regulatory functions toward a limited number of neighboring targets. Another possibility is that proteins, regulated by definite second messengers or enzymes, are susceptible only to a distinct pool of such secondary transducers.

Based on the above principles, three cardinal assumptions in relation to the mechanisms of intracellular parasitism of pathogenic microorganisms could be made. Firstly, due to the high accuracy of the signaling apparatus, the existence of effective mechanisms of autoregulation and the numerous collaterals for the flow of information, to block or affect signal transduction in eukaryotes could be a complicated task for intracellularly proliferating parasites. Secondly, since only a few types of secondary messengers transduce information inside the cell, bacteria possessing finite types of regulatory components able to synthesize, destroy or mimic such secondary transducers could affect virtually every pathway originating from any receptor. On the other hand, knowing which eukaryotic enzymes participate in signaling chains as key elements and by which modes these enzymes are regulated, one could predict which bacterial products will represent special danger to the normal operation of signaling cascades. And thirdly, to achieve their effect, regulatory products of bacterial origin must reach the distinct place of their action. They should be translocated to the site where their activity will produce the maximal, most profitable for the microorganism, effect.

Section II

Intracellular Parasitism of Bacteria

======= CHAPTER 11 =======

LEGIONELLA

GENERAL ASPECTS

The history of *Legionella* started in 1976, when a large outbreak of severe pneumonias occurred among participants at the Legionnaires' convention in Philadelphia, USA. During the following year the causative microorganism was isolated and termed *Legionella pneumophila*.[246] Later, additional species were identified and their number soon grew to more than 35. Half of these *Legionella* species are reported to be pathogenic for humans.

The source of the infection in most instances was aerosolized environmental water contaminated by *Legionella* . Thus the disease begins with inhalation of a sufficient dose of virulent bacteria. However strong evidence exists that consumption of drinking water or aspiration may also cause the infection.[268]

The predominant clinical manifestation of legionellosis is severe pneumonia (Legionnaires' disease). The other form of the disease is an acute self-limited nonpneumonic type termed Pontiac fever. The pathogenesis of the latter form is obscure. It has been proposed that Pontiac fever could be caused either by non-viable bacteria, by bacteria with reduced virulence or by *Legionella* engulfed in protozoa.

The pathological picture of Legionnaires' disease can be described as purulent pneumonia with lytic destruction of cells in the inflammatory infiltrate. Alveoli have been repeatedly shown to be filled with macrophages, neutrophils and fibrin. Large zones of hemorrhage were also a typical feature of *Legionella* pneumonia. The majority of bacteria could be seen within phagocytes in phagosomes and in the later phase, following extensive proliferation, within the cytosol of the cells.[400] Thus early pathological studies as well as the following investigations on virulence of *Legionella* clearly indicated the critical importance of intracellular multiplication of the bacterium for the development of full-featured Legionnaires' disease.

INTERACTION WITH PHAGOCYTES

The first step in *Legionella*-macrophage interaction is contact of bacteria with surface structures of phagocytes (Fig. 15). There is strong evidence that C3 receptors are involved in this process.[173] Recent data show however that these microorganisms are able to bind to certain cell lines, including alveolar macrophages, by a novel complement-independent binding mechanism as well.[142]

The binding of microorganisms to certain receptors might be the first attempt of bacteria to change proper signaling in the host cell. Efficient internalization of different bacteria by phagocytic cells through conventional routes was shown to result in stimulation of diverse antibacterial mechanisms in phagocytes. In contrast, ligation and stimulation of certain receptors by *L.pneumophila* may lead to generation of altered signaling cascade(s) resulting in inefficient oxidative burst and probably in other defects observed in antibacterial mechanisms of phagocytes.

Following receptor binding, unopsonized *L.pneumophila* were shown to be ingested into phagocytes by "coiling phagocytosis". In this process a long phagocyte pseudopods coiled around the bacterium as the microorganism was internalized. At the end of this process *L.pneumophila* cell stayed at the center of a large coil. In the following stages however the external portion of the coil disintegrated and the organism remained in a conventional membrane-bound phagosome.[172] In contrast, when *L.pneumophila* was coated with specific antibodies it was internalized by conventional phagocytosis. In this process tips of pseudopods met around the distal part of a microorganism and the bacterium was enclosed in a typical membrane-bound phagosome. Thus at the morphological level the results of coiling phagocytosis and conventional phagocytosis were the same. However the biochemical processes and signaling reactions triggered by both types of phagocytosis might be quite different.

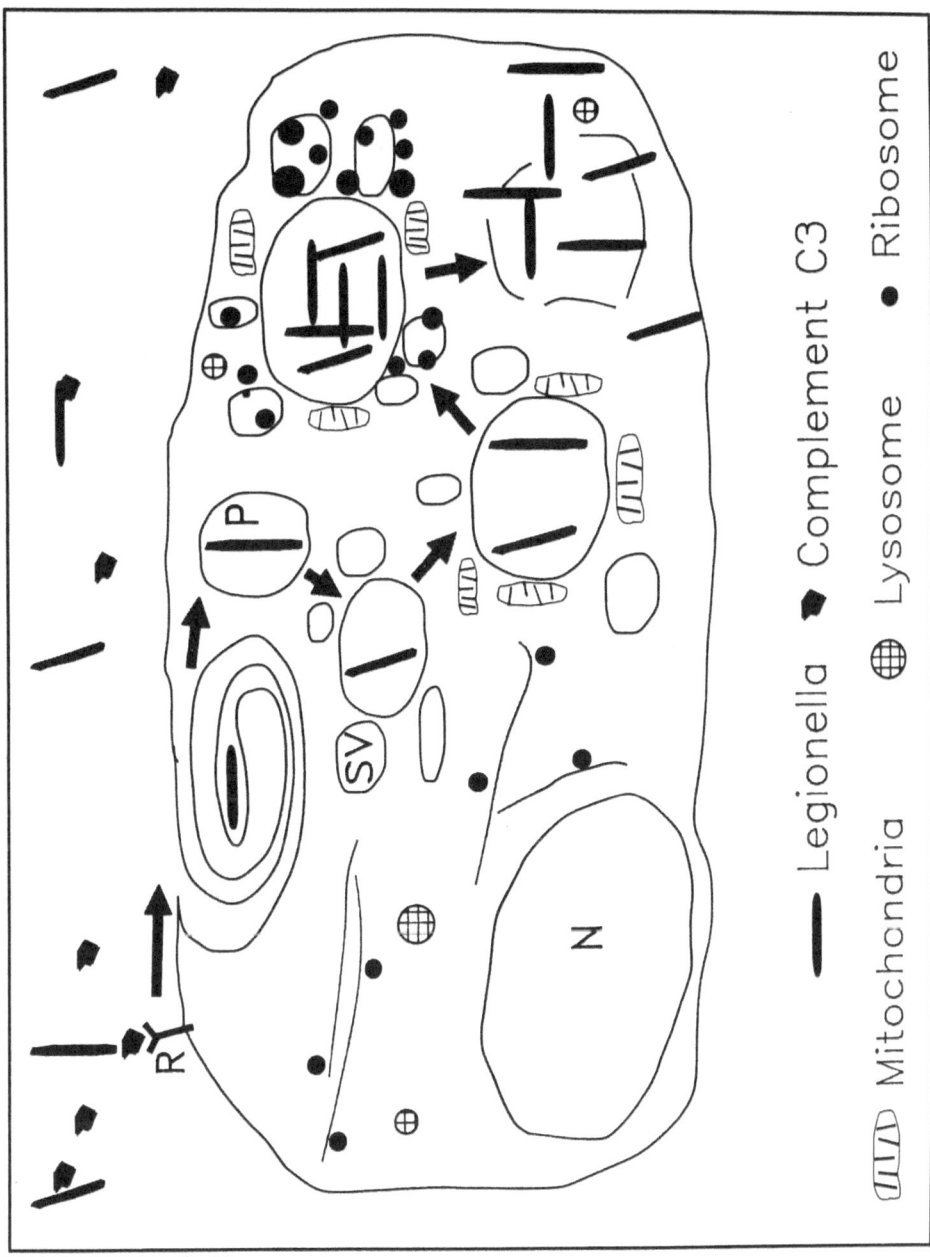

Fig. 15. Schematic view of Legionella-phagocyte interactions. R, receptor; SV, smooth vesicle; P, phagosome; N, nucleus. The binding of the microorganism is shown on this figure to be complement C3 receptor-dependent. However it is known that other types of receptors may participate in ligation of Legionella. For details see text.

— Legionella ➤ Complement C3 • Ribosome

〔//〕 Mitochondria ⊕ Lysosome

Studies on the composition of vacuole membrane during development and maturation of phagosomes demonstrated complex quantitative and qualitative rearrangements of its structure. Such changes in membrane composition are necessary for proper development of phagocytosis. In relation to *Legionella* it was shown that internalization of *L.pneumophila* via coiling or conventional phagocytosis resulted in exclusion of a vast majority of plasma membrane proteins from the phagosome,[85,86] including major histocompatibility complex I and II molecules and failure to accumulate transferrin receptor, CD63, LAMP-1 or LAMP-2.[87] One may speculate that these alterations in membrane composition could be caused by specific bacterial products, may result in altered signaling events and cause further defects in phagocytosis mechanisms for the benefit of *Legionella*.

Following internalization, the phagosome containing *Legionella* sequentially interacts with different cellular organelles, such as smooth vesicles, mitochondria and ribosomes. By 15 min phagosomes were shown to be surrounded by smooth vesicles. In the later phase, by approximately 1 hr, phagosomes were enclosed by mitochondria. And finally, by approximately 4 hrs following internalization, the bacterium-containing phagosomes were covered by ribosomes and ribosome-lined vesicles. Usually beginning with this phase an intensive multiplication of *L.pneumophila* could be observed.[170] The mechanism of these alterations and their importance both for *Legionella* and phagocyte biology remain a mystery.

The next prominent features of *Legionella*-phagocyte interaction include lack of phagosome-lysosome fusion and inhibition of phagosome acidification.[171,174] The mean pH value of phagosomes containing live virulent bacteria was shown to be consistently 0.8 units higher than the pH of phagosomes containing formalin-killed *Legionella*. The mechanism of defect in phagosome acidification may be linked to and be secondary to the inhibition of phagolysosome formation. However, the possibility exists that specific products of *L.pneumophila* may cause these defects separately.

Another abnormality in *Legionella*-phagocyte interaction, namely defective oxidative burst, has been described among others by Rajagopalan-Levasseur and colleagues.[311] These authors reported inhibition of the metabolic oxidative activity of polymorphonuclear leukocytes following ingestion of virulent *L.pneumophila*. The leukocytes displayed decreased rates of oxygen consumption and inhibition in chemiluminescence. In contrast, internalization of avirulent bacteria resulted in no inhibition of oxidative burst activity. These results suggested active inhibition of oxidative killing mechanisms accomplished by virulent *L.pneumophila*.

Thus there are several prominent features in *Legionella*-phagocyte interaction: (1) phagocytosis is mediated by specific receptors including those for complement components; (2) phagocytosis may proceed via a distinct coiling mechanism; (3) several eukaryotic membrane proteins are actively excluded from the phagosome containing *L.pneumophila*; (4) the phagosome retaining *L.pneumophila* interacts with and is surrounded by smooth vesicles, mitochondria and ribosomes; (5) fusion of phagosomes to lysosomes is inhibited; (6) acidification of *Legionella* phagosome is impaired; and (7) oxygen-dependent killing activity of phagocytes is inhibited. As a consequence of these described alterations, *L.pneumophila* cells are able to escape bactericidal action of phagocytes and are allowed to multiply in a phagosome. The above alterations are apparently triggered by specialized products of live *Legionella*, since killed bacterial cells are internalized by phagocytes in the usual manner and the resultant conventional phagosome does not interact with smooth vesicles, mitochondria or ribosomes; it normally fuses with lysosomes and is readily acidified.

The *Legionella*-induced defects in metabolism of phagocytes may reside at different levels. However one proposition is that alterations in proper signaling during phagocytosis produce incorrect regulation of complex biochemical events following bacterial uptake by the host cell and lead ultimately to inefficient killing of bacteria in phagocytic cells. These alterations may be caused by specialized *Legionella* products able to produce defects in signaling pathways of eukaryotic cells.

MAJOR OUTER MEMBRANE PROTEIN

The major outer membrane protein (MOMP, P29) of *L.pneumophila* is a cell wall component with a molecular mass of ca. 29 kDa.[130,167] In fact, there are slight variations in molecular mass of MOMP in this species in the range of 24-29 kDa.[281] The data

on distribution of major outer membrane protein in other representatives of *Legionella* genus is controversial. Some researchers observed a protein with similar molecular mass among different
species of *Legionella*,[59] whereas others described MOMP as
L.pneumophila-specific.[167] There is also a possibility that several
MOMPs exist in *Legionella*.

Using purified preparations, it was shown that P29 is a pepti-
doglycan-associated protein and that in the absence of reducing
agents the MOMP forms large complexes. In the presence of 2-mer-
captoethanol it dissociates into monomers.[130] In a special study
Gabay and colleagues demonstrated that the major outer membrane protein of *L.pneumophila* is a porin. The MOMP could be
readily reconstituted into planar black lipid membranes by promoting fusion of MOMP-containing vesicles with the membrane.
Investigating the specificity of *L.pneumophila* porin, it was shown
that it exhibited less selectivity for cations than anions and volt-
age-independent gating.

One probable participation of MOMP of *L.pneumophila* in
intracellular parasitism of the bacterium was speculated to involve
its porin activity.[173] It was proposed that *Legionella* may inhibit
acidification of phagosome, by inserting a proton ionophore (i.e.,
cation-selective MOMP) into the phagocyte membrane. However
the more likely role for P29 was suggested by the finding that
human complement component C3 was able to bind MOMP and
mediate C3 receptor-dependent phagocytosis.[19]

In a preliminary study it was shown that monocyte receptors
CR1 and CR3 for complement components C3b and iC3b respectively mediated phagocytosis of *L.pneumophila*.[296] In the subsequent experiments authors showed that bacteria fix C3 and this
fixation takes place by the alternative pathway of complement activation.[19] This study also demonstrated that C3 binds selectively
to MOMP on the *Legionella* surface, yielding a C3-acceptor molecule complex of ca. 150 kDa and another one of ca. 110 kDa.
C3 fixation might have been targeted to specific domain(s) of p29
molecule since cyanogen bromide-generated fragments bound
complement component differentially. To further characterize the
ability of *L.pneumophila* porin to fix C3 and mediate phagocytosis, MOMP was reconstituted into liposomes. The produced

MOMP-liposomes not only readily fixed C3 component, but once so opsonized they adhered to and were internalized by human monocytes.[19]

Thus, MOMP is a molecule that is able to direct phagocytosis through a specific pathway, which utilizes C3 components of complement and the corresponding phagocyte receptors. There is a possibility that this variant of phagocytosis might employ distinct signaling processes and cause at least some of the observed peculiarities of *Legionella*-phagocyte interaction.

CYTOLYSIN

One of the best studied cytotoxic factors identified in *Legionella* is cytolysin—a secreted protein which has also been termed Zn-metalloproteinase, major secretory protein or tissue-destructive protease.[24,38,89,108]

Legionella cytolysin is a 38 kDa protein [37] with proteolytic activity against different substrates, including collagen, gelatin, casein and chromogenic substrates Suc-O-Met-Arg-Pro-Tyr·pNA and Suc-Ala-Pro-Tyr·pNA. In addition, this *Legionella* product has been shown to inactivate interleukin-2, to cleave CD4 receptor and to split other different serum proteins.[264] Purified protease displays toxic effects against HeLa cells, fibroblasts, CHO cells and lysed erythrocytes. In animal models cytolysin could produce lesions resembling those of *Legionella* pneumonia, and when introduced in relatively large doses (50-100 μg per guinea pig) killed animals with a clinical picture of hemorrhagic lung edema.[89]

The *Legionella* protease has been demonstrated also to impair functions of phagocytes and natural killer cells. There was a report indicating that the purified *L. pneumophila* protease at non-toxic concentrations inhibited human neutrophil and monocyte chemotaxis toward various chemoattractants.[314] The delicate mechanism of this malfunction remains however unclear. In another study, concentrated supernatants from cultures of a virulent *L.pneumophila* strain inhibited spontaneous neutrophil chemotaxis, chemotaxis toward fMLP and O_2^- generation in response to zymosan-activated particles, PMA, A23187 and fMLP at concentrations that had no effect on cell viability.[326] Heat-treated supernatant from the virulent strain had no inhibitory effect on O_2^- generation in response to any of the four stimuli. In contrast to the wild-type strain, the supernatant from the protease-negative mutant failed to inhibit neutrophil O_2^- response to zymosan-activated particles and PMA and only partially inhibited neutrophil response to A23187 and fMLP. Neutrophil spontaneous migration was unaffected by the culture supernatant from the mutant, whereas directed chemotaxis was partially inhibited. These data suggested the importance of cytolysin in the impairment of diverse phagocyte functions. However, whether protease directly inhibits functions in phagocytic cells or participates in processing and activation of another biologically active molecule remains unanswered.

Bearing in mind the diverse activities of cytolysin we investigated its influence upon signaling reactions in eukaryotic cells, in particular on eukaryotic protein kinase cascades.[20]

Our preliminary experiments indicated that in the cytoplasmic fraction of lung tissue cells there exist one or more enzyme systems for phosphorylating proteins with a molecular mass of approximately 150 kDa and 55 kDa (Fig. 16). The addition of *Legionella* culture filtrate or purified cytolysin to the reaction system changed the autoradiographic picture: the 150 and 55 kDa phosphoproteins were no longer seen, but a phosphoprotein with molecular mass of 45 kDa was present. These results indicated that *Legionella* cytolysin took part in the change of the protein phosphorylation pattern seen in the cytosol of pulmonary cells. However it was not clear if the changes detected in the eukaryotic protein kinase system resulted because of splitting the acceptor proteins or because cytolysin serves as a polyfunctional enzyme with phosphokinase or phosphatase catalytic domains on a single molecule in addition to the proteolytic motif. To clarify the situation we partially purified the components of the protein kinase system for further investigation and showed that *Legionella* cytolysin specifically cleaved the 55 kDa acceptor protein of the phosphokinase system of lung cells to give a 45 kDa component. The absence of the 150 kDa protein may also be explained by the proteolytic action of cytolysin.

Our studies as well as those of other researchers indicated some importance of cytolysin in the pathogenesis of Legionnaires' disease. However investigations with mutant *Legionella* showed that

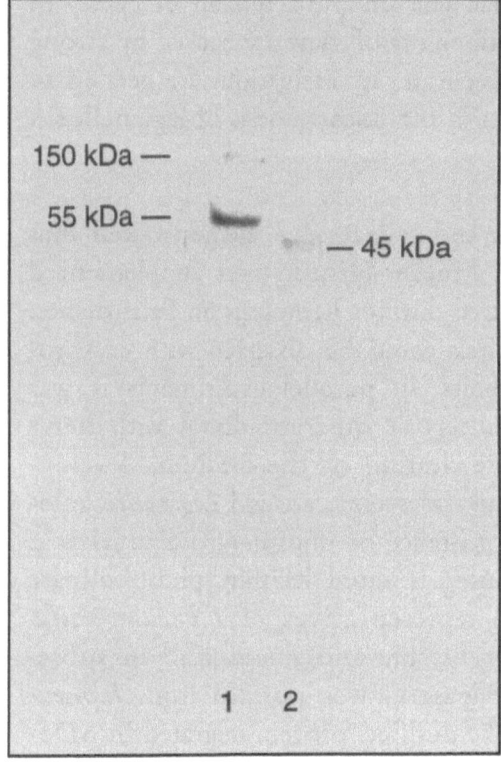

Fig. 16. Influence of Legionella cytolysin on protein kinase reaction in rabbit lung tissue cells. The reaction mixture, consisting of pulmonary cell cytosol, reaction buffer and [^{32}P] γ-ATP was incubated at 35 °C for 60 min without (line 1) or with (line 2) purified Legionella cytolysin. The components of the reaction mixture were then subjected to SDS-electrophoresis and autoradiography. Note the lack of phosphorylated components with molecular mass of ca. 150 kDa and 55 kDa and the appearance of a 45 kDa product as a result of metalloproteinase activity (line 2).

L.pneumophila possessing a specifically inactivated metalloproteinase gene was able to grow within and kill HL-60 cells, as did wild strain.[366] Moreover, guinea pigs challenged with the wild type and protease-negative mutant demonstrated comparable lethal doses, both strains multiplied at approximately similar rates and the histological picture of observed lesions was very similar.[39] It was concluded therefore that cytotoxic protease was required neither for intracellular infection or multiplication in phagocytes nor for lethality in laboratory animals.

Thus there is controversial data on participation of cytolysin in *Legionella* virulence and both points of view are backed by strong arguments. Additional, more delicate, investigations are needed to clarify the role of this product in the pathogenesis of legionellosis.

PHOSPHOLIPASE C

At the end of 1979 Baine and colleagues [15] demonstrated that growth of *L.pneumophila* on Mueller-Hinton agar supplemented with defibrinated blood produced distinct hemolysis on Petri dishes. The most pronounced hemolysis could be observed with erythrocytes of guinea pigs and rabbits. In parallel experiments it was shown that growth of *L.pneumophila* on Petri dishes with hen's egg yolk was accompanied by clearing of the medium, development of cloudy precipitate and iridescence around *Legionella* colonies. These data implied the existence of phospholipase activity in *Legionella* cultures. Indeed, using tritiated lecithin, phospholipase C activity was detected in representatives of *L.pneumophila*, *L.bozemanii*, *L.dumoffii*, *L.longbeachae* and *L.jordanis*.[13] In subsequent investigations phospholipase C was purified from *L.pneumophila* Dallas 1E strain.[14] The purified protein migrated in SDS-PAGE with an apparent molecular mass of 50-54 kDa. In contrast to the crude preparation, isolated enzyme failed to show hemolytic activity. Its phospholipase activity was maximal at basic pH (pH > 8.4) and could be enhanced by the presence of sorbitol and several nonionic detergents. SDS, EDTA, Cu^{2+}, Fe^{2+}, and Zn^{2+} inhibited the hydrolysis of substrate, whereas Ba^{2+}, Ca^{2+}, Co^{2+}, Mg^{2+} and Mn^{2+} restored the activity of phospholipase on EDTA-treated phospholipids.

The biological importance of *Legionella* phospholipase C is unknown. It did not lyse erythrocytes and was non-toxic for laboratory animals or hen's eggs. The only insight into the function of this product has been done using human neutrophils. In these preliminary experiments it was shown that phospholipase C purified from *L.pneumophila* inhibited neutrophil functions.[102]

Thus the above results indicate that several pathogenic *Legionella* species secrete phospholipase C. Potentially this enzyme could split lecithin, present in considerable quantities in eukaryotic cells, to produce a potent second messenger—diacylglycerol. Therefore the action of this *Legionella* product theoretically could influence eukaryotic signaling reactions, especially those that proceed via DAG, the protein kinase C pathway. However, detailed investigations on influencing cellular signaling machinery by *Legionella* phospholipase C have not been yet conducted.

PROTEIN KINASES

Data pointing toward the importance of protein phosphorylation in intracellular replication of *L.pneumophila* were obtained by Yamamoto and colleagues.[406] They demonstrated that infection of mouse peritoneal macrophages with a virulent *L.pneumophila* strain induced phosphorylation of a 76 kDa protein with unknown function(s) and this phosphorylation occurred at the early stages of *Legionella*-macrophage interaction.

In another study using mixed histones as a substrate, protein kinase activity was detected in crude sonicate of *L.micdadei*. The attempts to purify the corresponding enzyme(s) resulted in isolation of two proteins both possessing protein kinase activity, protein kinase 1 (PK1) and protein kinase 2 (PK2).[323] The two enzymes were characterized by similar pK value (0.27 and 0.29 mM respectively) and optimal pH (6.8-7.0). PK2 has been shown to be cAMP- and cGMP-dependent. Its activity could be stimulated by calmodulin, mixtures of Ca^{2+}-calmodulin, Ca^{2+}-phosphatidylserine, Ca^{2+}-phosphatidylinositol or Ca^{2+}-phosphatidylinositoldioleine. In contrast, PK1 has been shown to be cyclic-nucleotide-independent and failed to be stimulated by calmodulin or Ca^{2+} mixtures. PK1 was studied in more detail in the following experiments.[324]

Homogeneous PK1 was electrophoresed in SDS-PAGE as a single band with molecular mass of 55 kDa. This enzyme catalyzed the phosphorylation of several neutrophil proteins ranging in molecular mass from 11 to 38 kDa both in cytosol and membrane fractions of eukaryotic cells. In special experiments it was demonstrated that PK1 modified brain tubulin and phosphatidylinositol. In the latter case phosphatidylinositol-phosphate, which resulted from phosphatidylinositol, was not phosphorylated further to phosphatidylinositol-diphosphate. It was shown that following 4 hr of incubation with purified PK1, the level of phosphatidylinositol-phosphate was raised by 73-87%. In contrast, the level of phosphatidylinositol-diphosphate increased by not more than 10%.

Thus, *L.micdadei* contains at least two enzymes which can covalently modify numerous eukaryotic substrates. While protein kinase 2 is comparatively poorly studied, protein kinase 1 was investigated in more detail. In both cases however the precise biological function of the enzymes remains to be determined.

Among the physiological substrates for protein kinase 1 are tubulin and phosphatidylinositol. The modification of tubulin may result in defects in chemotaxis, altered formation of phagosomes and failure in phagosome-lysosome fusion. The phosphorylation of phosphatidylinositol is an intriguing phenomenon and may cause abnormalities in different signaling pathways of eukaryotic cells. For example, supra-optimal formation of phosphatidylinositol-phosphate in a cell may increase the concentration of this precursor of potential second messengers and thus exhaust normal signaling cascades and decrease cellular response to intruding microorganisms. In this connection it is interesting to note that another product of *L.micdadei*, acid phosphatase 2, is able to dephosphorylate phosphatidylinositol-diphosphate to phosphatidylinositol-phosphate. Thus, concerted action of both protein kinase and phosphatase could sharply elevate the level of phosphatidylinositol-phosphate. Another possibility is formation by *Legionella* protein kinase of an isomeric form of phosphatidylinositol phosphate (e.g., 3-isomer instead of 4-isomer) with different signaling activities. It should be noted also that the number of definite substrates of PK2 is unknown; therefore it is likely that activities of other eukaryotic proteins might be regulated by *Legionella* protein kinase as well. Another unanswered question is whether other species of *Legionella* in addition to *L.micdadei* synthesize similar enzymes.

PHOSPHATASES

Phosphatase activity was detected in *L.pneumophila* during both cytochemical and biochemical investigations in several laboratories.[171,269,282,377]

The attempts to purify this enzyme were performed by Saha and colleagues.[322] In this study two proteins with phosphatase activity were isolated from bacterial lysates. The acid phosphatase 1 (ACP1) has a molecular mass of approximately 150 kDa and failed to produce a discrete band on isoelectric focusing gels. The acid phosphatase 2 (ACP2) is an approximately 86 kDa enzyme with pI = 4.5. The substrates for ACP2 include among others phosphothreonine, phosphoserine and phosphotyrosine. Therefore this enzyme could influence eukaryotic signaling reaction by dephosphorylating specific proteins phosphorylated by host serine/threonine or tyrosine protein kinases.

The other feature of purified enzyme is its activity toward phosphatidylinositol 4,5-diphosphate and inositol triphosphate.[325] The reaction was specific in the former case since none of the phosphatidylinositol-4 phosphate formed was dephosphorylated further to phosphatidylinositol. Using neutrophils labeled with $^{32}P_i$, Saha and colleagues showed that by 30 min following addition of ACP2, 20% of the labeled phosphatidylinositol-diphosphate was dephosphorylated. When neutrophils were treated with phosphatase before their stimulation with fMLP, the concentration of inositol triphosphate and *sn*,-1,2-diacylglycerol was reduced by 44% and by 45% following the stimulation respectively. Thus the activity of *Legionella* phosphatase could influence generation of second messengers by eukaryotic phospholipases C through two proposed mechanisms.[102] First, the catalysis of phosphatidylinositol-diphosphate to phosphatidylinositol-phosphate left less phosphatidylinositol diphosphate as a substrate for phospholipase C in the host cell and increased the amount from phosphatidylinositol-phosphate possessing probably distinct signaling activities. Second, inositol triphosphate, formed by some activated phospholipase C, could then be readily dephosphorylated by phosphatase from *L.micdadei*. As a result, less inositol triphosphate is accumulated in a cell.

Studying the biological activities of purified *L.micdadei* phosphatases it was observed that preincubation of ACP2 with human neutrophils produced strong inhibition of O_2^- production following stimulation with fMLP and ConA,[102,322] though it had no effect on O_2^- production following activation with PMA.[102] In contrast to these results was a study by Kim and colleagues[201] who demonstrated that phosphatase-negative mutants and the parent strain of *L.pneumophila* were equally effective in their capacity to infect and grow within phagocytic cells.

Thus acid phosphatase 2 represents a product of *Legionella* which alters proper generation of the oxidative burst in phagocytes. It is very probable that the mechanism of such alteration includes influencing regulatory signaling reactions in the host cell by either dephosphorylation of second messengers in the route of phospholipid metabolism or by hydrolyzing phosphothreonine, phosphoserine or phosphotyrosine linkages in specific proteins, thus interfering with the action of host protein kinases.

ADP-RIBOSYLTRANSFERASE

In our initial work we obtained results indicating that ultrasonic lysates of *L.pneumophila* contained an enzyme that transferred the ADP-ribose moiety from NAD to several *Legionella* proteins. The ADP-ribosylating activity was strongly enhanced by the addition of crude macrophage lysates or the membrane fraction of macrophage lysates (Fig. 17) Several lines of evidence also indicated that one of ADP-ribosylated proteins (with molecular mass 54 kDa) was itself an ADP-ribosyltransferase which catalyzed the reaction of automodification.[22]

To characterize more precisely the observed phenomenon we developed a purification procedure to isolate the putative enzyme of *Legionella*.[23] However this task has not been completed successfully, since the material obtained after the last stage of purification of ADP-ribosyltransferase was not homogeneous. On SDS-PAGE gels we observed two major bands with molecular masses of 54 and 38 kDa as well as some minor components with molecular masses of 53 and 27 kDa (Fig. 18).

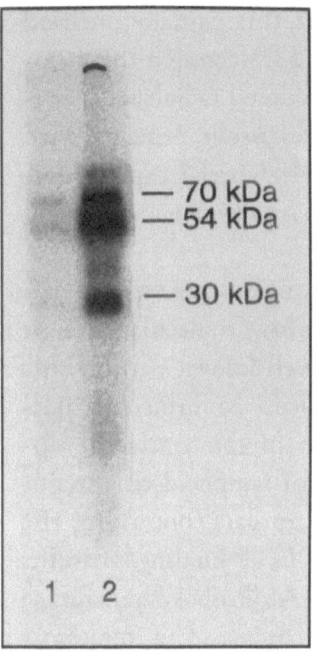

Fig. 17. Auto-ADP-ribosylation pattern of Legionella lysates. The reaction mixture, consisting of L. pneumophila ultrasonic lysate, reaction buffer and [^{32}P]NAD was incubated at 35 °C for 60 min without (line 1) or with (line 2) added macrophage lysates. After that the components of the reaction mixture were subjected to SDS-electrophoresis and autoradiography. Note the increased ADP-ribosylation of bacterial components in the presence of macrophage preparation (line 2).

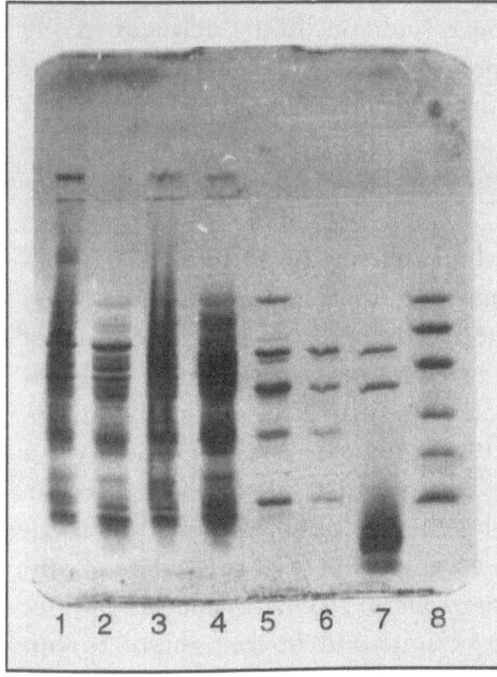

Fig. 18. SDS-electrophoresis analysis of ADP-ribosyltransferase of Legionella at different stages of purification. Ultrasonic lysate of L.pneumophila (line 1) was concentrated by ammonium sulfate treatment (line 2) and subjected sequentially to FPLC chromatography on Superose 6 column (line 3), DEAE-cellulose column (line 4), Mono Q column (line 5), Alkyl-Superose column (line 6) and Mono P column (line 7). Molecular mass markers (line 8) are from top to bottom 94, 67, 43, 30, 20.1 and 14.4 kDa. The final preparation (line 7) consisted of two major proteins with molecular mass of 54 and 38 kDa.

During the subsequent investigations of this partially purified preparation we observed that $MgCl_2$ and ATP decreased the auto-ADP-ribosylation reaction and strongly enhanced radiolabel incorporation into a group of rabbit lung cell cytosolic proteins with molecular masses of 20-25 kDa (Fig. 19). In special experiments we demonstrated that these targets of *L.pneumophila* enzyme were GTP-binding proteins.

Thus our data show that *Legionella* contains an enzyme capable of ADP-ribosylation of eukaryotic low molecular-weight GTP-binding proteins. Such proteins are well-known participants in signaling pathways in a cell and have been confirmed to participate in reactions as diverse as regulation of the contractile apparatus, oxygen burst generation, activation of lymphoid cells, regulation of cell growth etc. To date there are no data concerning the assignment of those low molecular weight GTP-binding proteins modified by the described *L.pneumophila* ADP-ribosyltransferase to some known group of GTP-binding proteins. This makes it impossible to discuss the biological function of this *Legionella* product in the host cell. However it should be noted that the ADP-ribosylation reaction was strongly and specifically enhanced in the presence of macrophage lysates. This suggests the possibility of activation of this enzyme during contact of *Legionella* with phagocytes.

MIP PROTEIN

In contrast to biochemical approaches to identify and characterize possible virulence factors, a group of researchers successfully used molecular-genetic strategies to study the pathogenesis of *Legionella* infection. In these investigations tools have been developed to induce and analyze in cell cultures various mutations in *L.pneumophila* genes.[83,115,297] It was disclosed that mutations in a gene coding for a surface 24 kDa protein resulted in severe reduction in virulence for macrophages, macrophage-like cell lines, alveolar epithelial cells and protozoa,[81,82,84] as well as leading to considerable attenuation of *L.pneumophila* in laboratory animals.[80] Since such mutants were shown to be impaired in their ability to initiate macrophage infection, the mutated surface component was named Mip (Macrophage infectivity potentiator) protein.

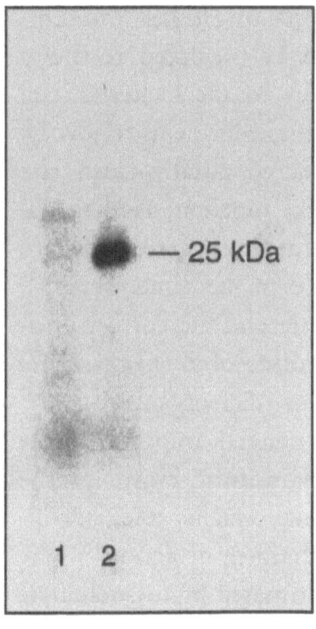

Fig. 19. ADP-ribosylation activity of purified Legionella enzyme. The reaction mixture, consisting of pulmonary cell cytosol, reaction buffer and $[^{32}P]NAD$ was incubated at 35 °C for 60 min without (line 1) or with (line 2) partially purified L.pneumophila ADP-ribosyltransferase. After that the components of the reaction mixture were subjected to SDS-electrophoresis and autoradiography. Note the strong ADP-ribosylation of 20-25 kDa eukaryotic proteins (line 2).

The deduced amino acid sequence of Mip protein from *L.pneumophila* [114] showed its homology to amino acid sequences of human, *Neurospora* and yeast FK506-binding proteins, which were able to bind the immunosuppressant drug FK506. FK506-binding proteins represent a receptor family of peptidyl-prolyl *cis/trans* isomerases (PPIs) termed immunophilins which are able to catalyze *cis/trans* interconversion of prolyl imidic peptide bonds in proteins.[217] Investigations of the 24 kDa Mip protein confirmed its isomerase activity. In addition, the inhibitory effect of FK506 on Mip was similar to that on human FK506-binding protein.[120]

The structural composition of Mip protein has been addressed in special studies. According to a hypothesis of Hacker and colleagues[154] the N-terminus of Mip, which is predicted to be a 60-amino acid α-helix, anchors the protein to the bacterial cell wall. The C-terminus, which carries the domain for peptidyl-prolyl *cis/trans* isomerase activity, might be projected distally from the bacterial surface to accomplish its biological function. Additional data using X-ray scattering for structure solution suggested that to perform its function Mip protein should be in the dimer state.[335]

Further studies showed that the *mip* gene and the corresponding protein could be detected in many strains of *L.pneumophila* and *L.micdadei* as well as in other intracellular microorganisms.[236,327] These data showed that Mip protein has a general, important role in the initiation of a cycle of intracellular parasitism. However the distinct mechanism of action of Mip protein remains unclear.

Among the speculations on Mip's mechanism of action[102] the first one we will consider is based on its estimated high isoelectric point (pI = 9.8). This dictates strong polycationic properties for the *Legionella* product. Polycations are known to induce phagocytosis of inert particles. Similarly, Mip protein could promote internalization of *Legionella* cells. Alternatively, the product of the *mip* gene could act as a cationic lysosomotropic agent, thus interfering with processes of phagosome acidification and phagosome-lysosome fusion. The other proposition is that Mip is a receptor that binds some regulatory product of phagocytes, thus diminishing its concentration in a cell. However the more probable explanation of Mip's mechanism of action could be linked to its enzymatic properties, e.g. peptidyl-prolyl *cis/trans* isomerase activity. The action of Mip protein may result in change of steric conformation of a target protein and affect its activity as a consequence. The role of target can be played by bacterial proteins or some eukaryotic products. The next possibility is that Mip, by analogy to eukaryotic immunophilins, may be a presenting molecule that is able to assist in the translocation of a certain bacterial or eukaryotic protein to the site in a signaling cascade where it can perform its activity. So, in the latter case after complexing with some molecule inside the phagocytic cell, *Legionella* Mip protein by analogy to FKBP or cyclophilin would be supposed to inhibit signaling events which lead to activation of phagocytic cells.

LOW-MOLECULAR WEIGHT TOXIN

The low-molecular weight cytotoxin was historically one of the first virulence factors purified from *Legionella* cultures. Originally it was identified in supernatants of *L.pneumophila* as a product cytotoxic toward CHO cells.[123] The activity was thermostable, methanol-soluble and stable at pH from 5 through 8. The cytotoxin was partially purified and was shown to represent a sextapeptide with a molecular mass of ca. 1200 Da. During investigations on its biological activities it was shown that cytotoxin influenced oxidative metabolism in polymorphonuclear phagocytes.[124,223] In concentrations which did not affect the neutrophil viability, the peptide suppressed the activity of hexosomonophosphate shunt (HMPS) activity and oxygen consumption during phagocytosis. Iodination and killing of *Escherichia coli* by neutrophils were also reduced. The effect of cytotoxin on HMPS activity was specific. Hexosomonophosphate shunt activity stimulated by A23187 and latex beads was reduced, while stimulation by ConA, PMA and valinomycin was unaffected. In another study cytotoxin was shown to also inhibit fMLP-induced O_2^- production.[102]

It is known that the fMLP-induced activation of O_2^- production proceeds as chain of reactions: fMLP receptor $\rightarrow$ G protein $\rightarrow$ phosphatidylinositol-specific phospholipase C $\rightarrow$ protein kinase C $\rightarrow$ NADPH oxidase. Since cytotoxin affected fMLP-induced stimulation of oxidative burst and failed to inhibit PMA (which is a specific activator of protein kinase C)-activated O_2^- production, the probable sites of its action should reside at steps before protein kinase C activation.[102]

In another study using bacterial ultrasonic lysates as a crude preparation, authors purified a *Legionella* lethal factor. This product was cytotoxic for macrophages, lethal to AKR/J mice, had molecular weight of 3400 Da and was present in several species of *Legionella*.[162] During investigations on its effect on neutrophil functions it was observed that purified toxin preparation significantly inhibited chemiluminescence of polymorphonuclear leukocytes, which is a marker of decreased oxidative burst activity.

Thus, two low molecular weight toxins have been purified from *Legionella* cultures. The precise mechanism of action of both moieties is not known, nor is it known if both peptides are identical molecules. Bearing in mind the very complex regulation of oxidative burst in phagocytes, the elucidation of the mechanism of toxins' action could be rather complicated. The toxic molecules were remarkably small, which almost certainly excludes the possibility of enzymatic activity. These toxins could act as "false" messengers in signaling cascades during regulation of oxygen metabolism, inactivate enzymes by competing for substrate or serve as "adapter" proteins in eukaryotic protein-protein interactions. Another possibility is that low molecular weight toxins represent ligands for some intracellular or surface receptor. However to date these propositions are truly speculative and need specific confirmation.

In a recent study Jacob and colleagues[188] demonstrated that *L.pneumophila* inhibited respiratory burst of monocytes by the downregulation of certain protein kinase C isotypes. It was shown that the phosphorylation of peptides with molecular mass of 34, 48, 62, 68 and 80 kDa stimulated by phorbol myristate acetate was markedly inhibited. In addition to this inhibition the expression of protein kinase C α and protein kinase C β isoforms was partially restricted. Whether such inhibition of protein kinase C activity has any connection with cytotoxic activity of *Legionella* cultures remains to be unanswered.

OTHER POTENTIAL REGULATORY PRODUCTS

The body of information concerning *Legionella* virulence and *Legionella*-phagocyte interaction is impressive. Principal features of the bacterium-host cell interaction have been well defined at the morphological level; however the mechanisms of observed alterations are unclear to date. In parallel, a panel of potential virulence factors of *Legionella* was isolated and characterized. However their role in the pathogenesis of Legionnaires' disease was obscure in most instances. This situation dictated the necessity for new approaches to be undertaken in order to delineate molecular mechanisms of *Legionella*-phagocyte interactions.

In the investigations described above authors have primarily used classic biochemical methodology: bacteria were grown in artificial media and the cultures were studied for the presence of products with specific biological or biochemical activities, e.g., cytotoxicity, lecithinase, proteinase activities etc.; the corresponding proteins were isolated and characterized. However it is clear that deeper knowledge about *Legionella* pathogenic mechanisms could be obtained from the experiments on expression of certain bacterial genes and on production of microbial proteins under the conditions of intracellular growth.

Thus, Rankin and Isberg[312] studied promoters regulated by the macrophage intracellular environment. The authors demonstrated that several dozens of proteins were apparently induced during intracellular multiplication of *L.pneumophila*. The results of that study were in good agreement with a paper by Abu-Kwaik and colleagues[2] who used another approach to investigate modulation of protein profiles during intracellular multiplication of *L.pneumophila*. Intracellular bacteria were radiolabelled and crude cell extracts were analyzed by two-dimensional SDS-PAGE. Authors showed that expression of at least 35 proteins was induced by intracellular environment, while at least 32 products were repressed intracellularly. In parallel experiments Miyamoto and colleagues[265] showed that several proteins were induced during intraphagocytic multiplication of the bacteria. These are novel products with molecular mass of 100, 65, 24 and 16 kDa. The functions of identified molecules are unknown to date.

Transposon mutagenesis was confirmed to be another effective tool to study genes and corresponding proteins necessary for intracellular proliferation of bacteria. A special enrichment protocol was developed to identify *L.pneumophila* transposon mutants specifically defective for intracellular growth.[25] On the basis of electron microscopy investigation, obtained mutants were divided into three main classes. The first one included bacteria which were able to avoid phagosome-lysosome fusion but displayed defects in forming phagosomes studded with ribosomes and ribosome-lined vesicles. The second group included *Legionella* that were unable to inhibit phagosome-lysosome fusion, in addition to showing defects in forming endoplasmic reticulum-encompassed phagosomes. The third class included members which showed delay in starting replication. The phenotype of the third class of mutants displayed features of previously described *mip* mutants. Mutations of the first two classes could be complemented by an open reading frame termed *dot* (defect in organelle trafficking) and coding for a 1048 amino acid residue DotA protein. In a special study it was confirmed that phenotypical alterations caused by the mutated gene were due to the disrupted *dotA* gene and not due to a polar effect on downstream genes.[26] Two general hypotheses were offered by the authors to explain the data obtained.[25,186] Firstly, the product of the *dot* gene could encode a positive regulatory factor which stimulated synthesis of different products responsible for organelle trafficking. Secondly, the *dot* locus could encode a single polyfunctional protein.

Brand and colleagues,[48] studying a mutated avirulent strain of *L.pneumophila,* identified a 12 kb fragment of bacterial chromosome which enabled the mutant *Legionella* cells to prevent phagolysosome fusion, to kill macrophages and to produce lethal pneumonia in guinea pigs. Authors termed this region the *icm* (intracellular multiplication) locus and demonstrated that it consisted of four adjacent genes *icm*WXYZ. The characterization of these genes and the corresponding proteins could supply a great deal of information concerning mechanisms of intracellular replication of *Legionella.*

CHAPTER 12

LISTERIA

GENERAL ASPECTS

Listeria monocytogenes represents a pathogen with a comparatively long record. The first indication of *Listeria* infection dates to 1926 when an outbreak of listeriosis was described in laboratory animals.[273] However only during the last two decades has the problem of listeriosis begun to attract the attention of the medical community, since large outbreaks and numerous sporadic cases of the disease have been documented not only in animal but also in human populations.

Among the known species of genus *Listeria* only *L.monocytogenes* and *L.ivanovii* appear to be pathogenic for humans. Both are grampositive, non-spore-forming bacteria which are able to replicate in a number of culture cells. The majority of human cases, however, were caused by *L.monocytogenes*, while *L.ivanovii* represents primarily an animal pathogen.

Listeriosis is an opportunistic infection and the newborn, elderly people and immunocompromised persons are the primary victims. The main clinical manifestations of human listeriosis include perinatal infections, abortions, meningo-encephalitis and sepsis. It is generally agreed that the principal route of transmission is consumption of foods contaminated by virulent *Listeria*. However other routes of transmission are supposed to exist, e.g., direct contact with infected animals or cross-infection during the neonatal period.[249,338]

Entry of *L.monocytogenes* into the host occurs normally in the gut. However what cells are used by the microorganisms as a port of entry are not known. Most probably the penetration of bacteria

into the host occurs via different cell types, presumably via epithelial cells and M cells covering the Peyer's patches.[342] Following
translocation through the intestinal barriers, *L.monocytogenes* are
phagocytosed by cells in the lamina propria.[310] Through the blood
stream the bacteria are then rapidly disseminated into different
organs, predominantly accumulating in liver and spleen. In these
organs *L.monocytogenes* is readily ingested by macrophages and the
larger part of the inoculum is destroyed.[90] However the survivors
begin to grow and are able to disseminate hematogenically to other
organs and tissues. Thus the phases of pathogenesis of listeriosis
are well investigated at a morphological level and demonstrate the
critical importance of intracellular replication of *Listeria* for the
progress of the disease.

INTERACTION WITH PHAGOCYTES

Pathogenic *Listeria* species are facultative intracellular parasites
well adapted for living in a variety of cells as well as extracellu-
larly. The intracellular cycle of *L.monocytogenes* consists of several
phases (Fig. 20).

Firstly, the microorganism is capable of attaching to a phagocytic cell membrane through complement components C1q and
C3b and the corresponding receptors. However as in the case with
another intracellular pathogen—*Legionella*—the participation of several other types of receptors is possible.[107]

Secondly, the prominent feature of *Listeria*-host cell interaction is an active induction of phagocytosis performed by the bacterium. The latter feature is most important for penetrating into
the non-professional phagocytes, such as parenchymal, endothelial
or epithelial cells. The bacteria are internalized in a membrane-
bound vacuole which is typically lyzed during the following
20-30 min.[132] Molecular analysis of this step showed that at least
two proteins are involved in the process of internalization of
L.monocytogenes by phagocytes, a 60 kDa major extracellular protein p60 and internalin, a product of the *inlAB* gene.[131,215]

The InlA protein has molecular mass of 88 kDa and is attached
to the listerial cell wall via its C-terminal domain. It includes two
repeat regions. The N-terminal 15-fold repeats are composed of
strings of 22 amino acid residues and are highly homologous to a

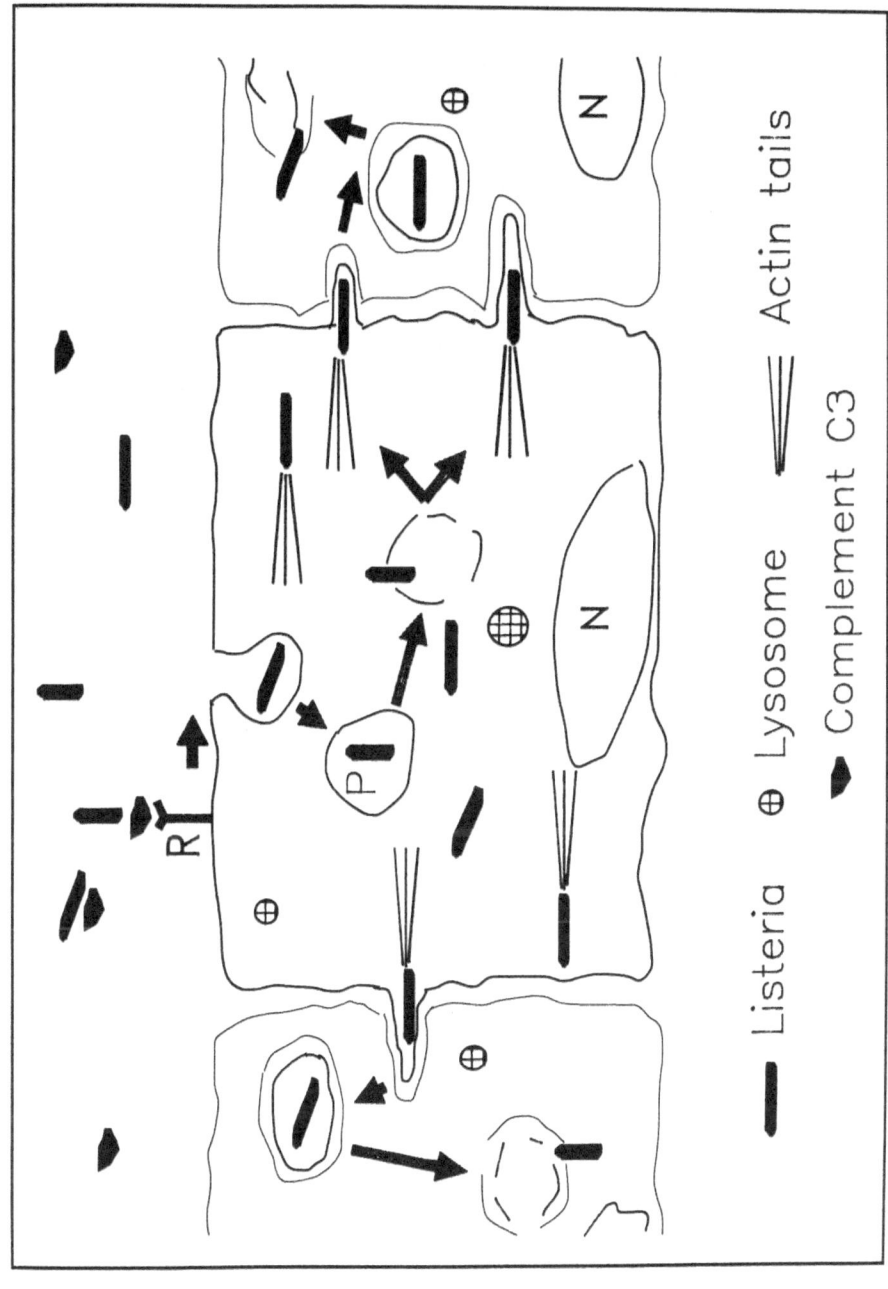

Fig. 20. Schematic view of Listeria-phagocyte interactions. R, receptor; P, phagosome; N, nucleus. Three adjacent cells are shown. For details see text.

— Listeria   ⊕ Lysosome

⋙ Actin tails

≪ Complement C3

superfamily of leucine-rich proteins (e.g., YopM of *Yersinia pestis* and hemagglutinin FHA of *Bordetella pertussis*). The C-terminal repeats are homologous to surface protein M of *Streptococcus pyogenes* and presumably represent a spacer for the functional exposing of the first region of repeats.

InlB is a 65 kDa cell wall-associated protein. Recent study by Dramsi and colleagues[106] demonstrated that InlB was required for entry of *L.monocytogenes* into hepatocyte cells and some hepatocyte cell lines but not into intestinal cell lines. These results provided evidence that differential expression of InlA and InlB might confer specificity to *Listeria* cell tropism. The N-terminal part of InlB is composed of repeats with analogy to InlA, while the C-terminus is highly hydrophobic and functions to attach the whole protein to the listerial surface. The importance of internalin genes in invasive processes of *Listeria* was confirmed by site-directed mutagenesis. Mutations in the *inlA* gene resulted in considerable reduction of bacterial infectivity, while deletions in the *inlB* gene resolved in almost total reduction of invasive potential. Thus, both proteins are necessary for full invasiveness of *L.monocytogenes*. However, complementation analysis showed that neither InlA nor InlB alone or in combination were sufficient to provide invasiveness to a non-pathogenic strain of *L.innocua*.[65] Results of another study[230] demonstrated that internalization of *L.monocytogenes* by dendritic cells was apparently independent of the *inl* locus. These data show that yet unidentified additional products of *Listeria* participate in the process of bacterial invasion.

Another substance which is important for invasion by *Listeria* is the p60 major extracellular protein. This product represents murein hydrolase, which is required for the cell division process. Thus p60 is of vital importance for the *Listeria* life circle and can be found in all tested strains of the genus. However special experiments showed that a mutant *L.monocytogenes* strain in addition to altered morphology displayed decreased adherence to and invasiveness into cell lines.[215] Pretreatment of this mutant with purified p60 restored bacterial cell and colony morphology and augmented *Listeria* virulence. Moreover, an attenuated recombinant *Salmonella typhimurium* strain expressing p60 was shown to display increased invasion of hepatocytes and macrophages.[166] To explain the mechanism of participation by p60 in the invasive process the following propositions based on sequence composition of this protein were done. One idea was that positively charged amino acid residues of the molecule improve the interaction between the bacterium and eukaryotic cell via electrostatic force. Another possibility is that p60 participates in proper organization of the listerial cell wall in such a manner that certain proteins (among others internalin) are allowed to be expressed correctly and act properly to mediate phagocytosis.

The third step in *L.monocytogenes*-host cell interaction includes lysing of host vacuoles and free replication of the bacteria in the cytoplasm. This step is dependent primarily on listeriolysin O, a 58 kDa thiol-dependent hemolysin.[139] It was shown that mutants lacking functionally active listeriolysin O and the wild strain were equally effective in entering cell lines; however the former failed to lyse phagosomes and thus were unable to replicate in the cytoplasm. The virulence of mutated strains for laboratory animals was also severely reduced.[132] The low pH found in phagosomes favored the maximal activity of listeriolysin O since its lytic potential was highest at acidic pH. In another study,[138] attenuated *S.dublin* strain secreting functionally active listeriolysin was partially released into the cytoplasm of J774 cells, in contrast to a nonhemolytic strain which remained in the phagosome.

Subsequent investigations demonstrated however that listeriolysin O is not the only factor responsible for lysing of *Listeria*-containing vacuoles. Nonhemolytic mutants of *L.monocytogenes* were able to grow inside certain cell lines. This fact pointed to the possibility of other lysing factors.[304] A candidate for such a factor, which might act in concert with listeriolysin O to lyse the membrane of primary vacuoles, was phosphatidylinositol-specific phospholipase C (PlcA).[224,252]

It was shown that deletion in a gene encoding PlcA affected escape of *L.monocytogenes* from the phagosome.[61] Growth of such a mutant in infected mice was also reduced. These investigations indicated the possibility of direct participation by PlcA in lysing of eukaryotic membrane, probably in cooperation with listeriolysin O. However the narrow substrate specificity of this phospholipase made it difficult to explain the membranolytic activity of PlcA by its direct enzymatic action. The recent study of Goldfine and colleagues[146] shed some light on this problem. It was demonstrated that purified PlcA induced leakage from liposomes which was independent of phospholipid hydrolysis. In control experiments another phospholipase, phosphatidylinositol-specific phospholipase from *Bacillus thuringiensis*, readily bound to membrane but failed to induce liposome lysis. Moreover, at comparatively low concentrations PlcA and listeriolysin O, added separately, could induce slow membrane leakage. Simultaneous addition of both products resulted in rapid membrane permeabilization with additive effect. The delicate mechanism of pore formation by listerial PlcA is unknown. Possible mechanisms may include: (1) formation of pores due to oligomerization of PlcA molecules at the membrane surface and insertion of such oligomers into the bilayer by analogy to streptolysin O;[32] (2) insertion of amphipathic regions of PlcA into lipid membrane; and (3) formation of pores due to multiple interactions of charged protein groups with lipid polar head groups.[146]

The fourth step in *Listeria*-phagocyte interaction represents movement of the bacterium from one host cell into another. This movement has been shown to depend on the ability of *L. monocytogenes* to induce polymerization of eukaryotic host actin into "clouds" and "comet tails". Such actin "clouds" can be observed around *Listeria* and are subsequently organized into the "comet tail" located at one pole of the bacterial cell.[378] The continuous polymerization of actin produces rapid movement (ca. 1-1.5 µm/s) of bacteria, thus propelling *Listeria* forwards. At the end of this stage the bacterium reaches the host cytoplasmic membrane and, moving ahead, becomes incorporated into a "fingerlike" protrusion into a neighboring cell which is then internalized. To identify bacterial factors participating in movement of *Listeria*, transposon-induced mutants defective in cell-to-cell spread have been isolated and characterized.[209] It was disclosed that a certain protein (ActA) is involved in observed actin reorganizations.

ActA is a surface protein. Based on nucleotide sequence, its molecular mass was calculated to be 67 kDa and its pI = 4.74. The protein has been shown to be rich in glutamic acid and proline residues. The hydrophobic C-terminus is thought to participate in anchoring of the protein to the bacterial membrane and is followed by a positively charged tail. The anomalous mobility of ActA during SDS-electrophoresis (the calculated $M_r = 90$ kDa) is apparently caused by this tail. The ActA molecule contains two long proline-rich repeats and a small motif repeated seven-fold. Two thirds of the protein are speculated to be projected outside the bacterium and are supposed to interact with eukaryotic cytoskeletal machinery to achieve actin nucleation.[65]

The mechanism by which ActA stimulates actin polymerization is not known in full detail and is being intensively studied. ActA is located asymmetrically on the bacterial cell during division; it is absent in one end (the "new-born" end) and is abundant in the other (the "old" end). In the cytoplasm of infected cells the protein was found to be localized to the site of actin filament formation and could not be located in the "comet" tail.[210,412] These observations suggest that ActA triggers the process of actin polymerization at one pole of the bacterium, possibly interacting with some eukaryotic protein(s). One evidence of such interaction resulted from the observation that ActA is phosphorylated by some eukaryotic protein kinase.[55] Another protein which was shown to directly interact with ActA (in particular with its unphosphorylated form) represents VASP, vasodilator stimulated phosphoprotein.[109] Studying mutated forms of ActA it was shown that the region between amino acid residues 128 to 151 represents an important element for actin filament nucleation.[65] Deletion in the actin-nucleation region of ActA resulted in a recombinant non-motile *L.monocytogenes* strain capable of binding VASP. In contrast, deletion in the proline-rich region produced a strain able to accumulate actin but incapable of producing actin tails. This phenotype correlates with an absence of binding VASP.[65] Thus at least two domains are presented on ActA molecules: a nucleation domain required for actin binding and a VASP-binding domain necessary for actin dynamic rearrangement.

VASP was first described in eukaryotes as a protein which was located near the cytoskeleton structures and became phosphorylated by protein kinase A and protein kinase G in response to vasodilator agents. As was shown in *Listeria* research,[109] entry of bacteria into the cytoplasm results in rapid accumulation of VASP at the pole proximal to the actin tail. Another protein which was able to bind VASP and participate in actin tail elongation was profilin, which may accelerate the process of adenine nucleotide exchange, thereby promoting ADP-actin to react with ATP to form ATP-actin and thus allowing actin polymerization.[351] Thus the proposed scenario of VASP participation in listerial movement could be as follows: VASP binds the unphosphorylated form of ActA and recruits certain proteins (among them profilin, vitamin D-binding protein etc.). These proteins interact with actin filaments located near the nucleation site of ActA and directly participate in dynamic elongation of the growing actin tail. Phosphorylation of ActA performed by some unknown protein kinase produces conformational change in the molecule, with subsequent inability to bind VASP. Thus only bacteria-associated, freshly made molecules of ActA bind VASP, whereas older forms of ActA are left in phosphorylated form in the wake of the moving bacterium and do not interfere with VASP-ActA interaction.

In addition to ActA, certain yet unknown listerial proteins are apparently needed for proper actin reorganization. This proposition arose from the detection of a mutant *L.monocytogenes* strain which was severely affected in its capacity to form "comet tails" but was able to induce actin polymerization.[342]

The last stage of *Listeria*-host cell interaction is lysis of the double membrane of a phagosome. This type of phagosome is produced by invagination of the cytoplasmic membrane into the new host cell as a result of listerial propelling. Listeriolysin O is a factor which apparently participates in this process. However, bearing in mind the different composition of a double-membrane phagosome in contrast to a primary phagosome (outside-in in a primary vacuole vs. inside-out in a secondary vacuole), listeriolysin O faces membrane with altered lipid contents (thus distribution of cholesterol molecules, which are receptors for listeriolysin O, might be different). Therefore, participation of accessory listerial products is required. Such an accessory protein which is necessary for lysing of secondary vacuoles is broad range or phosphadidylcholine-specific phospholipase C (PlcB).[388]

Phosphatidylcholine-specific phospholipase C of *L.monocytogenes* is a 29 kDa enzyme which hydrolyzes a broad range of glycerophospholipids and sphingomyelin[140,144] in membranes. Mutants of *L.monocytogenes* lacking the production of an active PlcB produced considerably smaller plaques on fibroblast monolayers than did the wild strain.[388] During electron microscopic studies it was shown that the obtained mutant strain accumulated in secondary double-membrane vacuoles. As a consequence, a decreased number of bacteria could be found free in cytoplasm of host cells.

As described above, PlcA and PlcB play specific roles at different stages of *Listeria*-host cell interaction. However, more detailed studies demonstrated some overlapping in their activities. Thus strains mutated in both *plcA* and *plcB* genes were considerably more impaired in their escape from the primary vacuole than were *plcA* mutants. Analogously, defects in lysing of secondary vacuoles during cell-to-cell spread were more prominent in double mutants than in *plcB*- only deficient strains.[143]

The interaction of *Listeria* with host cells has been studied in considerable detail. All stages of intracellular replication of the bacterium are well documented and include binding of *Listeria* to specific receptor on the surface of eukaryotic cells, activation of phagocytosis, lysing of a primary vacuole, free replication in the cytoplasm of a host cell, active modification of cytoskeletal components to organize actin tails necessary for penetration of *Listeria* into the neighboring cell, lysing of a double-membrane secondary vacuole and free multiplication in the cytoplasm of the neighboring cell. This cycle can be repeated many times and allows the bacteria to invade host tissues. The molecular mechanisms of the described processes have been intensively studied and may represent the basis for explanation of the described stages in pathogenesis of listerial infection. However, among the virulence factors of *L.monocytogenes* known to exist several are clearly capable of eliciting signaling functions. These proteins will be described in more detail in the following sections.

PHOSPHATIDYLINOSITOL-SPECIFIC PHOSPHOLIPASE C

A gene coding for phosphatidylinositol-specific phospholipase C (PlcA) was detected by comparison of a sequence of an open reading frame ORF U of *L.monocytogenes* with that of data bank entries.[252] This gene bank search revealed homology of investigated listerial sequence with phosphatidylinositol-specific phospholipases C of *B.cereus* and *B.thuringiensis*. The *plcA* gene encoded a 317 amino acid peptide with molecular mass of 33 kDa. In experiments on detection of phosphatidylinositol-specific phospholipase C activity, *L.monocytogenes* cultures revealed the presence of phospholipase, primarily in the supernatant fluid.

During subsequent studies PlcA was purified to homogeneity and characterized.[145] It is a strongly basic protein with pI = 9.4. In crude liquid culture it formed complexes with anionic proteins and migrated during gel-chromatography as a high molecular weight complex. The investigation of substrate specificity of purified enzyme displayed no activity on phosphatidylethanolamine, phosphatidylserine, phosphatidylcholine, phosphatidylinositol-phosphate and phosphatidylinositol-diphosphate. The main substrates of PlcA include phosphatidylinositol and phosphatidylinositol-glycan.

The role of PlcA in virulence of *L.monocytogenes* was demonstrated by studying the effect of transposon mutations in the *plcA* gene.[60,61] It was demonstrated that these mutants were severely impaired in their virulence, tested in a mouse model, and failed to grow in cell cultures. Electron microscopic studies demonstrated defective escape of mutant bacteria from phagosomes of macrophages as compared to the parent strain of *L.monocytogens*.

In another study[340] it was shown that PlcA, introduced into the non-pathogenic *L.innocua* strain, conferred some virulence potential to this microbe. Thus macrophage-like J774 cells cocultivated with the wild strain of *L.innocua* ingested and killed the bacteria, with falling viable counts during the first 8 hr post-infection. In contrast, recombinant *L.innocua* strain secreting an active enzyme was similarly phagocytosed by these cells while showing a different growth curve. The viable counts rose during 6-8 hr post-infection with subsequent decline during the following 16 hr. Wild-type *Listeria* could be found as single cells residing in phagosomes, whereas an *L.innocua* strain expressing PlcA was able to begin replication in vacuoles which were normally acidified and had undergone phagosome-lysosome fusion. These results demonstrated again that this phospholipase is important for intracellular replication of *Listeria*. However in contrast to previous studies, this importance was demonstrated to be at least partially independent of its pore-forming activity. In other words, PlcA of *L.monocytogenes* may in addition to pore formation play another role in intracellular replication of *Listeria*.

As mentioned above, phosphatidylinositol-specific phospholipase C demonstrated considerable activity against glycosyl-phosphatidylinositol. It was speculated therefore that one mode of PlcA participation in bacterial virulence could be removal from membranes of certain biologically active proteins (e.g., receptors, enzymes) anchored through GPI. However special studies demonstrated almost no activity of PlcA in releasing the proteins anchored by a GPI moiety to certain cell types.[133] For example, listerial phospholipase in contrast to phospholipase C from *B.thuringiensis* failed to split off GPI-anchored acetylcholinesterase from erythrocytes and cleaved only a low amount of alkaline phosphatase from kidney cells or Fcγ receptor from granulocytes. These results demonstrated that GPI-anchored proteins of eukaryotic origin may not be the natural substrates for phosphatidylinositol-specific phospholipase C of *L.monocytogenes*.

Attempts to investigate the effects of PlcA enzymatic activity on the signaling machinery of eukaryotic cells have been performed recently. The obtained experimental data showed that J774 cells infected with virulent *L.monocytogenes* displayed considerable increase in diacylglycerol contents compared to double mutants in PlcA and PlcB.[143] A major effect of this second messenger is activation of protein kinase C. Therefore one could propose that intracellular replication of *Listeria* might be accompanied by alterations in signal transduction caused by hyperactivation of certain protein kinase C isoforms.

In accordance, special investigations confirmed the importance of protein phosphorylation reactions in the intracellular parasitism of *Listera*. For example Tang and colleagues[370] demonstrated that during invasion of epithelial cells by *L.monocytogenes* certain isoforms of MAP kinase became tyrosine phosphorylated. The tyrosine kinase inhibitor genistein blocked both bacterial uptake and modification of these proteins. In another study[389] authors demonstrated the inhibition of virulent *Listeria* entry into epithelial cells accomplished by several specific inhibitors of tyrosine protein kinase activity. Moreover the phagocytosis of *L.monocytogenes* and

L.ivanovii into host cells could be restricted by different inhibitors. It was shown that *L.monocytogenes* entry was decreased by erbstatin, genistein and some tyrphostins. In contrast entry of *L.ivanovii* was not affected by genistein and tyrphostins.

These data demonstrated the complex character of protein modifications in phagocytic cells as well as the importance of protein phosphorylation reactions in phagocytosis of *Listeria*. One of the protein kinases which probably plays a central part in regulation of processes following internalization of *Listeria* is protein kinase C. This enzyme has been shown to regulate a vast number of cross-talk reactions in eukaryotic cells, among them coordinated activity of tyrosine protein kinases (such as receptor tyrosine kinases and MEK which is a MAP kinase kinase). Therefore, the influence of DAG produced by PlcA on protein kinase C activity may result in altered chains of tyrosine phosphorylation reactions and, as a consequence, modified signal-driven processes in a host. In addition to its effect on the MAP kinase cascade, protein kinase C has been demonstrated to directly participate in coordination of cytoskeletal rearrangements during phagocytosis, regulation of respiratory burst and other bactericidal events. Therefore increased production of DAG by listerial phospholipase may result in severe malfunctions in eukaryotic cell metabolism and thus apparently favor bacterial multiplication in the infected host.

Other products of phosphatidylinositol-specific phospholipase C of *L.monocytogenes* are apparently inositol phosphate produced from phosphatidylinositol and phosphoinositol-glycan generated from GPI. The role of these metabolites produced by listerial enzyme in intracellular signaling reactions during replication of *Listeria* has not been studied. Among the inositol second messengers the most important product is apparently inositol 1,4,5-triphosphate generated from phosphatidylinositol-4,5-diphosphate. One of the main functions of this molecule is regulation of Ca^{2+} mobilization. However it was demonstrated that various products of IP_3 metabolism are potent signaling molecules as well and perform specialized yet insufficiently studied functions in cellular metabolism. Such products include inositol phosphates with different numbers of phosphate groups: inositol 1,4- or inositol 1,5-diphosphate, (cl:2,4,5)IP$_3$, (3,4,5,6)IP$_4$ and (1,3,4,6)IP$_4$, IP$_5$ and possibly also IP$_6$. The complex interconversions of these substances are regulated by specialized enzymes—kinases and phosphatases. Therefore an increase in IP contents caused by listerial PlcA may cause shifts in the equilibrium of different inositol phosphates and might alter the normal operation of signaling cascades in eukaryotic cells.

Inositol-glycan messengers have been proposed to participate in many signaling reactions (e.g., insulin response, events following ligation of Ly-6A.2 and Thy-1 receptors etc.). One hypothesis describing action of inositol-glycans in signal transduction considers several steps in the generation of this transducer. Firstly, extracellular proteinase removes protein from protein-GPI complex. This process triggers internalization of GPI attached to cytoplasmic membrane. Secondly, specific intracellular phospholipase C degrades GPI to release inositol-glycan second messenger which forwards the signal downstream. In relation to listerial virulence one could speculate that the function of such intracellular eukaryotic phospholipase could be accomplished by phosphatidylinositol-specific phospholipase C of *L.monocytogenes*. As a result, abnormal production of inositol-glycan could favor intracellular survival of *Listeria* in one way or another.

PHOSPHATIDYLCHOLINE-SPECIFIC PHOSPHOLIPASE C

Phosphatidylcholine-specific phospholipase C (PlcB) was purified from liquid culture of *L.monocytogenes*. The active enzyme represented a 29 kDa protein stimulated by Zn^{2+} ions. PlcB has been shown to be produced as proenzyme with molecular mass of 33 kDa which is converted into the mature product by listerial metalloproteinase. Testing various strains of *Listeria* in immunoblot with monospecific serum, authors demonstrated the occurrence of PlcB only in strains of *L.monocytogenes*. The purified PlcB was active at neutral and weakly acidic pH, possessed little hemolytic activity and was non-toxic for mice. The exoenzyme hydrolyzed a broad spectrum of phospholipids, including phosphatidylcholine, phosphatidylethanolamine, phosphatidylserine and sphingomyelin, but failed to hydrolyze phosphatidylinositol.[140] During the following more detailed study on enzymatic activity authors demonstrated weak activity of PlcB on phosphatidylinositol, cardiolipin, plasmalogens and some other lipids present in mixed micelles.[144]

The cloning of the *plcB* gene demonstrated the homology of this 289 amino-acid protein to phosphatidylcholine-specific phospholipases C from *B.cereus* and *Clostridium perfringens*.[388] Mutants defective in PlcB activity were produced and characterized. One type of such mutants had transposon insertion into the metalloproteinase gene, which is necessary for proper maturation of PlcB.[101,313] By immunoblot analysis it was demonstrated that the loss of lecithinase activity was paralleled by the absence of an active 29 kDa enzyme and the presence of an inactive 33 kDa form. Through the analysis of *plcB* insertion mutants, evidence has been obtained that PlcB contributes to lysis of the double membrane of the secondary vacuoles. Electron microscopic study demonstrated that such bacteria resided in vacuoles and possessed decreased capability to lyse phagocytic membranes. Virulence of PlcB-deficient strains has been shown to be reduced and the bacteria produce small plaques on fibroblast monolayers.

The contribution of phosphatidylcholine-specific phospholipase C of *L.monocytogenes* to the virulence of bacteria has been discussed mainly from the point of view of its membranolytic activity. However eukaryotic counterparts of PlcB are well known to have clearly established regulatory functions. This fact points to the possibility of participation of listerial PlcB in signaling reactions as well.

The products of enzymatic activity of PlcB apparently may include DAG and phosphocholine, phosphoethanolamine or phosphoserine originating from phosphatidylcholine, phosphatidylethanolamine and phosphatidylserine respectively. As described above, DAG is a potent second messenger and is involved in activation of protein kinase C. Another biologically-active product of PlcB, ceramide, may be produced from the cleavage of sphingomyelin. It is generally assumed that products of sphingomyelin hydrolysis represent second messengers with inhibitory activity toward different signaling reactions. Firstly, they may repress phosphorylation of different substrates by protein kinases. Thus enzymatic activity of PlcB, probably in concert with that of PlcA, hypothetically may accomplish regulation of cellular machinery while producing potent second messengers. Ceramide arising from the hydrolysis of sphingomyelin may repress one subfamily of protein kinases, while DAG generated from phosphatidylcholine and other triglycerides could apparently activate another group of phosphorylating enzymes. The balance between stimulatory and inhibitory stimuli accomplished by synthesized secondary messengers may result in fine manipulation of eukaryotic metabolism.

LISTERIOLYSIN O

Listeriolysin O (HlyA) is a potent hemolysin produced by *L.monocytogenes*. Two other species, namely *L.ivanovii* and *L.seeligeri*, secrete very similar hemolysins termed ivanolysin and seeligerolysin.[98,223] Listeriolysin O is apparently the best characterized product of *L.monocytogenes* and its role in virulence of *Listeria* is thought to be well established.

The gene coding for listeriolysin O was cloned and sequenced.[253,254,392] The *hlyA* determinant encodes a protein of 504 amino acid residues in its secreted form and directs synthesis of a mature product with a molecular mass of 58 kDa. The signal sequence consists of 25 amino acid residues.

Listeriolysin O has been purified from listerial cultures as an extracellular product with a molecular mass of ca. 60 kDa.[139] This protein possesses all the classical properties of SH-activated hemolysins: inhibition by cholesterol, activation by reducing agents such as 2-mercaptoethanol or dithiothreitol, and antigenic cross-reactivity with the prototype of this class of hemolysins—streptolysin O of *Streptococcus* spp.

Listeriolysin O, comparably to other thiol-dependent hemolysins, has a narrow pH optimum. Thus its activity was shown to be highest at pH = 5.5 and almost no hemolysis was observed at pH = 7.0 and above.[139] In addition to high hemolytic activity (ca. 10^6 hemolytic units per 1 mg of purified protein) against sheep, human, horse and rabbit erythrocytes, HlyA possessed a lethal effect on experimental animals with LD_{50} of ca. 0.8 µg. Mice inoculated with purified protein died with convulsions and opisthotonus. Introduced intradermally, hemolysin in a dose of as low as 10 ng caused rapid local inflammatory response.

The mechanism of lysis of erythrocyte membranes by listeriolysin O has not been studied in detail but appears to be similar to that of streptolysin O.[32] That is, water soluble molecules of hemolysin bind to cholesterol-containing target membranes and are assembled into curved rod structures made of hemolysin oligomers. These structures form rings and arcs that penetrate the bilayer. The convex face of the rod structures are hydrophobic, in contrast to a hydrophilic concave surface. The embedding of such structures generates large transmembrane pores that allow free transfer of molecules.

Critical role of listeriolysin O in listerial virulence has been confirmed in investigations on non-hemolytic mutants of *L.monocytogenes*.[95,196,304] In electron microscopy investigations it was demonstrated that mutated strains are effective in entering eukaryotic cells but remain in phagosomal vacuoles and thus fail to gain access to cytoplasm for effective replication.[132] Concomitantly with decreased replication in host cells, severe reduction of virulence (ca. 10[5]-fold) has been observed. When the gene coding for HlyA was introduced into such mutants, the virulence was restored. In another study *hlyA* was cloned into the non-invasive pathogen *B.subtilis*.[33] Following expression of listeriolysin O, this strain lysed the phagosomal membrane and grew in the macrophage cytoplasm.

These results demonstrated that listeriolysin O is an important virulence factor of *L.monocytogenes* and participates in lysing of vacuoles (both primary and secondary) of host cells, thus allowing free multiplication of the bacteria in the cytoplasm. However recent data point to the possibility of participation of listeriolysin O in altering signaling reactions in host cells as well.

To study the effects of listeriolysin O on eukaryotic cells the authors used a novel approach. They transfected a bacterial gene for HlyA of *L.monocytogenes* into fibroblast and epithelial cell lines and studied the results of its expression in the cytoplasm of these cells.[98] In these experiments *hlyA* genes with and without signal sequence were introduced into 3T6, 3T3 and L2 cells under the control of the metallothionein promoter, which allowed specific activation of the corresponding genes by $ZnSO_4$. It was shown that 3T6 and L2 cells transfected with an intact listeriolysin O gene but grown without $ZnSO_4$ displayed little if any alteration in growth rate or cellular morphology. In contrast, stimulation of *hlyA* expression resulted in dramatic changes in both characteristics. Under the conditions used, large cell aggregates which consisted of roundup cells were formed. Such enhanced proliferative activity was paralleled by increased formation of actin microfilaments. When the expression of a leaderless form of listeriolysin O was stimulated, 3T6 cells were subjected to cytopathic changes while L2 cells were unaffected.

Inasmuch as enhanced proliferation and focus formation were observed with listeriolysin O containing signal sequence and since HlyA added to the medium of eukaryotic cells produced no proliferative effect, it was postulated that the bacterial product entered the secretory pathway of eukaryotic cells and reached the inside surface of a cytoplasmic membrane. Listeriolysin O is secreted by *L.monocytogenes* residing in a phagosome and it could be inserted into the vacuole membrane and presumably could be recycled with the rest of membrane fragments. The other origin of functionally active listeriolysin O might be *L.monocytogenes* replicating free in the cytoplasm of the infected host. In both cases, when introduced into the cytoplasmic membrane, HlyA could trigger some signals which regulate cellular functions, among them cell proliferation. One outcome of such increased proliferative activity may be generation of a larger number of cells available for invasion by *Listeria*. In contrast, leaderless listeriolysin O remains in the cytoplasm of host cells and produces cytopathic effects, presumably due to trivial lysing of membranous apparatus.[98]

While the described phenomenon is very interesting and displays a new area of listeriolysin O activity, the fine mechanism of observed defects remains unknown. One proposition is that the HlyA molecule disturbs the composition of the inner surface of the cytoplasmic membrane, changes the interposition of signaling structures of host cells and stimulates abnormal signaling pathways. Future studies are necessary to clarify the mechanisms involved in morphological defects observed in transfected fibroblast and epithelial cell lines.

OTHER POTENTIAL REGULATORY PRODUCTS

The biology of intracellular parasitism of *Listeria* is relatively well investigated. The main products of the microorganism which are believed to be critical virulence factors are defined and their role in the pathogenesis of listeriosis appears to be clear. However it is also obvious that the described genes and virulence factors of *Listeria* do not explain the entire complex of *Listeria*-host cell interaction in full detail.

The many functions of relatively well characterized listerial products remain a mystery. It is not clear how internalin contributes to the entry of bacterial cells and what receptors on the surface of the host organisms are necessary for its binding. The exact mechanism of p60 action is also not known; does this protein represent the ligand for some unknown eukaryotic receptor or does it participate in proper modeling of the listerial cell wall, thus contributing to the correct structural organization of other surface factors? It is also not known what signaling pathways are activated following contact of these listerial products with host cell receptors.

The functional importance of phosphatidylinositol-specific phospholipase C is also not fully defined. If PlcA possesses pore-forming activity which is apparently independent of hydrolysis of phosphatidylinositol, what is the exact role of phosphatidylinositol-specific phospholipase C activity in the pathogenesis of the disease? Both listerial phospholipases, PlcA and PlcB, are capable of producing potent second messengers following degradation of certain phospholipids (phosphatidylinositol, phosphatidylcholine, sphingomyelin etc.). The result of increased levels of such products on eukaryotic metabolism is completely unstudied.

Listeria incorporates eukaryotic cytoskeletal machinery for intra- and intercellular motility. However the exact mechanism of participation of ActA protein in this process remains in many details uninvestigated. It is known that actin rearrangement is a very complex process which is regulated by numerous signaling reactions with direct participation of Ras proteins of the Rho family, several protein kinases, Ca^{2+} ions etc. The role, if any, of ActA in triggering such regulatory pathways awaits further study.

It is certain that further investigations into *L.monocytogenes* will uncover new aspects of functions of known bacterial products. One example of such re-investigation is a recent study on listeriolysin O which displayed its capability to initiate complex signaling processes in eukaryotic cells.

Special studies have demonstrated that the number of known virulence factors of *L.monocytogenes* represents only part of the possible bacterial products important for intracellular parasitism of *Listeria* and the pathogenesis of listeriosis. Therefore, thorough investigations into the molecular biology of *L.monocytogenes* will certainly result in isolation of new biologically active substances.

Certain unidentified proteins have been observed by comparison in SDS-electrophoresis of listerial strains with different virulence.[367] Another approach which could be fruitful for elucidating novel virulence determinants is investigation of the molecular microbiology of stress conditions in *Listeria*. Using this approach several products specifically induced under the conditions of heat, cold or nutritional shock have been identified.[41,121,225,302,349] Their impact on the virulence of *Listeria* awaits further study.

An intriguing observation has been done[342] demonstrating the existence in the *Listeria* genome of a sequence(s) which produced a strong hybridization signal with a gene for cholera toxin. The potent regulatory activity of cholera toxin and its capability to modulate numerous eukaryotic processes is well known. Thus it is very interesting to investigate the impact of products of such a listerial gene in intracellular parasitism of bacteria and in pathogenesis of listeriosis.

===================== CHAPTER 13 =====================

MYCOBACTERIUM

GENERAL ASPECTS

Representatives of *Mycobacterium* spp. are gram positive bacilli of generally 1-4 x 0.3-1 μm in size and widely distributed in the environment. Members of the genus *Mycobacterium* are very diverse in their biology and clinical importance. Most of them appear to be saprophytic inhabitants of surface water and the soil. However, few are important human and animal pathogens. The most important species is apparently *M.tuberculosis*. These mycobacteria represent infectious agents of tuberculosis, one of the global problems for medicine. Epidemiological data suggest that there are ca. 1 billion persons infected in the world with 8 million new cases and 3 million deaths annually.[363] The next medically important species, *M.leprae*, causes leprosy, a chronic systemic disease affecting more than 10 million people in the world.[50] The last group includes "atypical" *Mycobacteria* also termed "Mycobacteria other than *M.tuberculosis*". The diseases caused by the latter group are beginning to emerge and in some human populations the incidence of these pathogens surpasses that of *M.tuberculosis*.

The microorganisms of the "atypical" *Mycobacterium* group differ from *M.tuberculosis* in that they do not cause person-to-person transmission, their species are ubiquitous in nature and they may colonize either human or animal without causing the disease. These mycobacteria are opportunistic pathogens and some predisposing conditions (e.g., physical trauma, malignancy, AIDS etc.) are needed for an infection to start. Among the potentially pathogenic species of "atypical" mycobacteria are *M.avium-intracellulare* complex, *M.kansasi*, *M.scrofulaceum*, *M.fortuitum-chelonae*, *M.xenopi*,

M.ulcerans and some others. The spectrum of the clinical manifestations is also diverse. Thus *M.avium-intracellulare, M.kansasi* and
M.xenopi may cause pulmonary disease which sometimes resembles
typical tuberculosis. *M.scrofulaceum* largely causes lymphadenitis and
rarely pneumonia, while *M.fortuitum-chelonae* and *M.ulcerans* are
infectious agents of skin, soft tissues and the musculo-skeletal
system.[403]

   In spite of rather diverse epidemiological, microbiological and
clinical aspects all pathogenic mycobacteria possess a determinant
critical for their virulence. That is, *M.tuberculosis* and other pathogenic mycobacteria can survive and replicate in a wide number of
phagocytic cells. Such multiplication inside phagocytes (primarily
in monocytes and macrophages) represents a key step in the pathogenesis of corresponding diseases. This aspect has been best studied in the cases of *M.tuberculosis* and *M.avium-intracellulare*.

INTERACTION WITH PHAGOCYTES

   The initial step in the interaction of *Mycobacterium* with phagocytes is contact of the microorganisms with surface receptors of a
eukaryotic cell (Fig. 21). As in the case with many other intracellular parasites (e.g., *Legionella* or *Listeria*) one probable group of
receptors for binding of *Mycobacterium* are receptors for complement C3.[333] This binding triggers the process of internalization of
a parasite which proceeds till total engulfment of microorganisms
into phagosomes occurs. Another type of receptor which is important in the process of internalization, apparently only for virulent
M.tuberculosis strains, is mannose receptors.[334] Thus mannose re-
ceptor-specific binding could be the first attempt of the *Mycobacterium* to change the proper course of phagocytosis and to avoid
generation of a full-featured bactericidal attack in the eukaryotic
cell. In addition to C3 complement receptors and mannosyl-fucosyl
receptors, fibronectin receptors have been also shown to be important in binding and uptake of *M.avium*.[28]

   One of the early malfunctions observed during the studies on
Mycobacterium-phagocyte interaction was altered sorting of
phagosome membrane proteins.[87] It is known that phagosomes
containing particulate material should undergo several steps in
maturation events before they fuse with lysosomes for complete

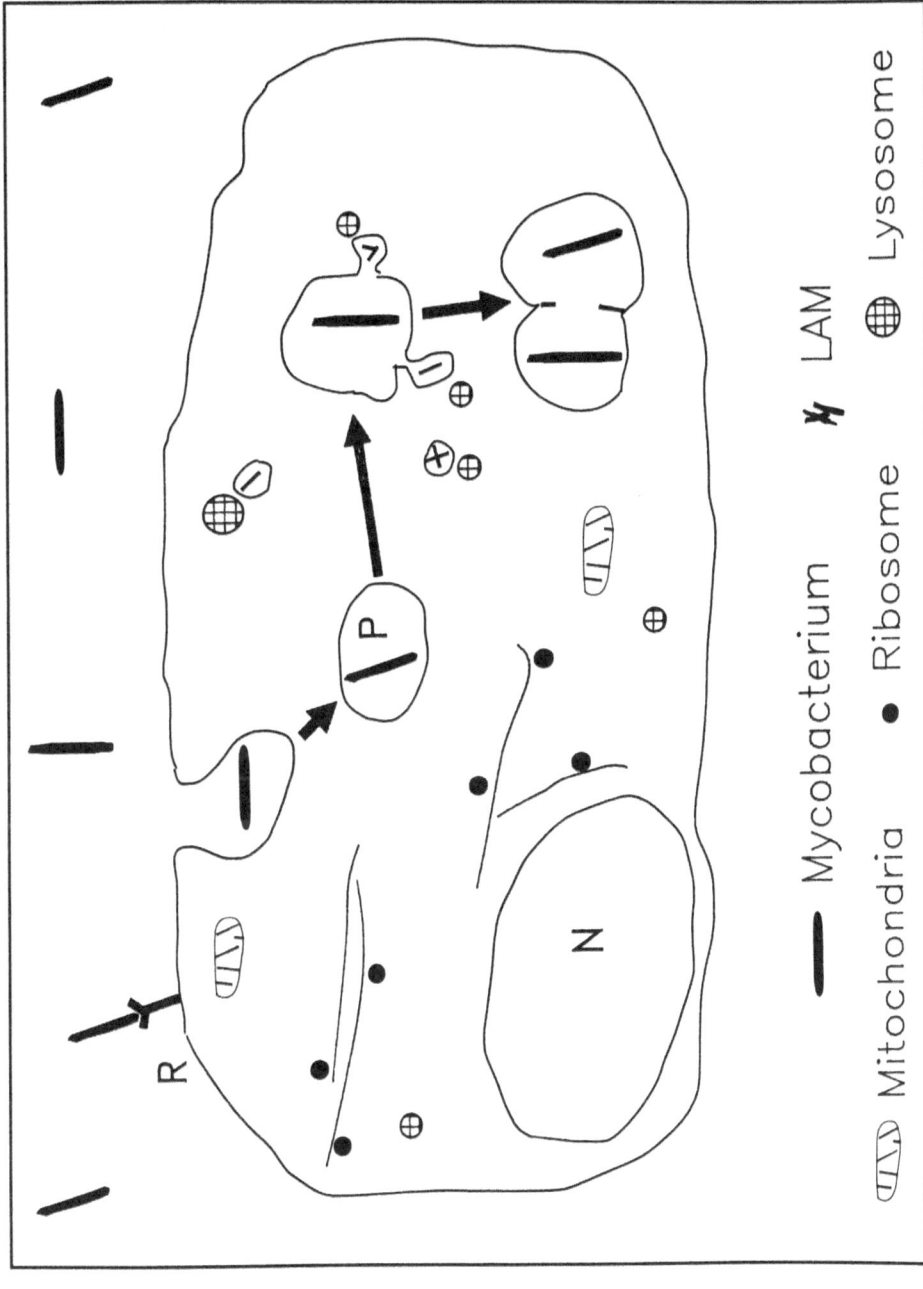

Fig. 21. Schematic view of Mycobacterium-phagocyte interactions. R, receptor; P, phagosome; N, nucleus. For details see text.

degradation of their contents. Primary phagosomes have been shown to normally lose several endosomal markers while acquiring markers associated with a late endosome, and finally to fuse with lysosomes to acquire lysosomal hydrolases and specific membrane glycoproteins. In relation to maturation of *M.tuberculosis*-containing phagosomes it was demonstrated that such phagosomes failed to exclude MHC class I and class II molecules as well as the known endosomal marker transferin. The decreased maturation of *M.tuberculosis*-containing phagosomes was also demonstrated by markedly diminished amounts of lysosomal markers cathepsin D and specific membrane glycoproteins.[87]

Inhibition of phago-lysosome fusion was another alteration in the normal course of phagocytosis frequently seen during *Mycobacterium*-phagocyte interaction.[11,87] For instance, phagosomes containing live tubercle bacilli showed very few fusion events with BSA-gold labeled lysosomes following 3 hrs-5 days post infection in contrast to either killed mycobacteria or inert particles. In agreement with these data mycobacterial phagosomes have been demonstrated to contain low levels of lysosomal marker proteins CD63, LAMP-1, LAMP-2 and cathepsin D.[87] As one mechanism of decreased fusion ability of lysosomes observed after phagocytosis of virulent *Mycobacterium*, the inhibition of lysosomal movements has been proposed.[159]

Another probable alteration in phagocytosis of *M.tuberculosis* has been described by McDonough and colleagues.[247] These authors demonstrated that both virulent and avirulent strains of mycobacteria were readily phagocytosed and allowed the phago-lysosome fusion to proceed. However in the later stage, virulent mycobacteria appeared to bud out from these fused phagosomes into vacuoles which did not fuse with secondary lysosomes. The following multiplication of tubercle bacilli occurred in these vacuoles. In addition to mycobacteria which replicated in vacuoles, a considerable number of *M.tuberculosis* were able to escape phagosomes and multiply free in the cytoplasm. In contrast, the avirulent BCG strain did not exhibit such escape behavior and failed to replicate to significant numbers.

One characteristic feature of mycobacterial multiplication is vesicular traffic out of mycobacterium-containing phagosomes. Such "daughter" vacuoles contain considerable amounts of a lipid product of *Mycobacterium*, lipoarabinomannan (LAM), and readily fuse with lysosomes.[404] One conjectural outcome of this phenomenon could be exhaustion of the bactericidal machinery of a phagocytic cell which lost its degrading enzymes by hunting "wrong targets".

Inhibition of acidification of mycobacterial phagosomes is another alteration widely illustrated in the literature. It has also been demonstrated that crude protein preparations from *M.tuberculosis* cultures, when added to macrophage-like cell line J774A.1, produced concentration-dependent and saturable alkalization of lysosomal contents.[74] The authors suggested that some mycobacterial antigen(s) or toxin exists which is capable of affecting the function of the lysosomal apparatus participating in acidification of lysosomes.

In subsequent investigations Sturgill-Koszycki and colleagues demonstrated that the relative alkalization of mycobacterial phagosomes might be caused by an active exclusion of the proton-ATPase.[362] These investigators showed that phagosomes with *M.avium* never acidified below pH = 6.3-6.5. Immunoelectron investigation of macrophages containing *Mycobacterium* demonstrated the presence of LAMP-1, which is a lysosomal-endosomal marker protein. In contrast, antibodies against E and B subunits of the mammalian vesicular proton-ATPase as well as against a 110 kDa ATPase accessory protein failed to detect the corresponding products in the vacuoles containing mycobacteria.

The phagosome containing *Mycobacterium* is subjected to replenishment of membrane which is lost during budding-out of LAM-containing vesicles or dividing of mycobacteria. Yet some membrane components (e.g., LAMP-1) are sustained whereas others (e.g., proton-ATPase) are excluded. These data demonstrate that membrane replenishment in the case of mycobacteria is a selective process and such specificity is actively upheld by the replicating microorganism. It is unclear however what bacterial product(s) participates in this process and what signaling pathways are affected.

Another malfunction of the antibacterial apparatus often observed in phagocytes containing mycobacteria is diminished activity of respiratory burst in eukaryotic cells. Gordon and Hart demonstrated that *M.microti* and *M.bovis* BCG generated low

respiratory burst as compared to zymosan.[147] The administration
of zymosan to macrophages following mycobacterial infection induced less oxygen burst than zymosan without *M.microti* or BCG.
The authors suggested that while zymosan-induced respiratory burst
was of surface origin, the inhibition of zymosan-induced respiratory burst by preexposure of phagocytes to mycobacteria could
originate from the intracellular space and was apparently caused
by specific product(s) of invading microorganisms.

Despite the immense difficulties in investigation of *Mycobacterium* (e.g., rather slow growth, problems with gene transfer experiments etc.) the intracellular biology of mycobacteria and their
interaction with phagocytic cells are studied in considerable detail.
Virulent microorganisms have been shown to induce multiple defects in macrophages—inhibition of phagosome-lysosome fusion,
pinching-off of vesicles containing lipoarabinomannan, decrease of
phagosome acidification and reduced generation of respiratory burst.
There is also data about possible escape of mycobacteria from
phagosomes fused with lysosomes and their free replication in the
cytoplasm. The information on the latter subject is however controversial and the majority of investigators believe that these microorganisms multiply in phagosomes which are able to divide as
the number of mycobacteria rises.

These accumulated data demonstrate that representatives of
Mycobacterium are able to affect the complex antibacterial machinery of phagocytic cells. One probable cause of such alterations is
inducing changes into the regulation of bactericidal processes following engulfment of microorganisms. It is tempting to speculate
that specific products of bacterial cells could induce the observed
malfunctions by interfering with signaling events in eukaryotic cells.

LIPOARABINOMANNAN

One of the best studied products of pathogenic *Mycobacterium*
is lipoarabinomannan, a component of the mycobacterial cell wall.
Early studies on the native product demonstrated that LAM consists of glycerol, inositol, phosphate group, arabinose, mannose,
lactate, succinate, palmitate and tuberculostearate.[50] The detailed
structural analysis suggested that LAM is a phosphatidylinositol-
glycan anchored in the mycobacterial cell wall by two fatty acid

residues linked to a glycerol moiety at the first and second positions. In the third position glycerol residue there is a phosphoester linkage with inositol which is associated with extensively glycosylated side chains. The terminal ends of LAM consist of branched or linear arabinofuranosides capped in the case of *M.tuberculosis* by mannose residues (the structure termed ManLAM). The molecules devoid of end mannose residues are termed AraLAM. In the latter case the constructions could be partially capped with phosphoinositol residues.

LAM exhibits a large spectrum of biological activities and is supposed to represent one of the main virulence factors of pathogenic mycobacteria.

Early studies on the mechanisms of binding of *Mycobacterium* to macrophages indicated that one of the receptors participating in adherence of virulent strains of *M.tuberculosis* is mannose receptors.[332,390] In a follow-up study it was demonstrated that the terminal mannosyl residues of ManLAM isolated from the virulent Erdman strain of *M.tuberculosis* conferred adherence of polystyrene microspheres to human macrophages.[334] In contrast, LAM from an avirulent strain (AraLAM) and lipomannan failed to promote binding of modified particles. The critical role of ManLAM in binding of mycobacteria to macrophages was supported further using monoclonal antibodies against LAM or competitive assay with purified LAM. Antibodies as well as LAM preparation consistently inhibited the adherence of virulent *M.tuberculosis* to human macrophages.[334,358]

Following ingestion of mycobacteria, lipoarabinomannan has been documented to be produced and secreted in considerable quantities. Based on this observation this compound has been suggested to play a scavenger role for bactericidal substances produced by macrophages. However more detailed investigations demonstrated pleiotropic effects of LAM on very different aspects of phagocytic cell physiology.

It was shown that LAM purified from a virulent strain of *M.tuberculosis* failed to elicit TNF-α secretion either alone or in combination with IFN-γ, did not increase the level of transcription of c-*foc* and production of chemoattractants KC and JE in murine macrophages. In contrast, LAM prepared from avirulent strains elicited TNF-α secretion and induced expression of c-*foc*, KC and JE. It is interesting to note that LAM from the virulent Erdman strain was able to synergize with IFN-γ for induction of JE. Since the latter product is chemotactic for monocytes, the initiation of its synthesis might stimulate recruitment of mononuclear phagocytes which are potential targets for subsequent colonization by mycobacteria.[318,319]

In another study Chan and colleagues[69] demonstrated that LAM purified from *M. tuberculosis* inhibited protein kinase C activity and blocked the transcriptional activation of IFN-inducible genes in macrophage-like cell lines. For example, purified preparation of lipoarabinomannan produced 80% inhibition of protein kinase C activity at a concentration of 270 μM. To study the effect of LAM on IFN-γ-inducible genes, authors investigated the expression of γ.1 and the major histocompatibility gene HLA-DRβ. In both cases strong inhibition was observed. Removal of acyl groups of LAM completely abolished its inhibitory ability.

It is known that protein kinase C is a critical component of a vast number of signaling pathways including respiratory burst activation, cytoskeletal rearrangement and intracellular vesicular traffic, and is an important component in mediating the effects of several cytokines (e.g., TNF-α, IFN-γ etc.). Therefore the influence of this protein kinase activity might be one of the central mechanisms in observed LAM-induced alterations in macrophages.

Despite the fact that the array of effects accomplished by lipoarabinomannan from *M. tuberculosis* is relatively well studied, the exact mechanism of its action is not defined. Since a molecule of LAM contains a phosphatidylinositol moiety, one could propose that lipoarabinomannan may serve as a precursor in the production of lipid second messengers inositol-phosphate and diacylglycerol. The breakdown of LAM inside the host cell could be accomplished by some phospholipase of either mycobacterial or macrophage origin. DAG produced from LAM is a hydrophobic molecule and is hypothesized to insert into the membranous structures of eukaryotic cells where it can contact protein kinase C. Since the composition of acyl residues of LAM differs from that of eukaryotic phosphatidylinositols in that it contains mycobacteria-specific tuberculostearate (10-methyloctadecanoate), the effect of LAM-derived DAG on protein kinase C might be opposite to that normally observed with eukaryotic DAG (i.e., inhibition versus activation).

Another proposition partially confirmed by experimental data is that mycobacterial LAM, following its excretion from the mycobacterial cell wall, becomes incorporated into the cytoplasmic membrane of the target cell and interferes with signal transduction mechanisms involving eukaryotic GPI-anchored receptors, e.g., Thy-1.[185] In this process the unsplit molecule of LAM apparently plays its role since intact acyl chains and mannoside glycan have both been shown to be necessary for specific integration into the cytoplasmic membrane.

PROTEIN KINASE

Protein tyrosine phosphorylation has been described in *M.tuberculosis* cultures by Chow and colleagues.[79] Using anti-phosphotyrosine monoclonal antibodies they detected one major protein migrating at 55 kDa in three *M.tuberculosis* strains tested. Since many tyrosine kinases are normally subjected to auto-phosphorylation, the observed protein might be protein kinase as well. During screening of several strains of *Mycobacterium* for the presence of tyrosine-phosphorylated protein(s) only *M.tubercolosis* has been shown to contain a 55 kDa modified protein. In contrast, *M.smegmatis*, *M.fortuitum* and *M.intracellulare* failed to contain proteins identified by the monoclonal antibodies used.

Protein tyrosine kinases are the principal components of numerous signaling pathways and in many examples they participate in initiation of signal transduction by modifying certain proteins at the early stages of signaling cascades. A considerable body of information is present which documents the cooperation of protein tyrosine kinases (e.g., protein kinase of the Src family) in the signaling events leading to activation or inhibition of leukocytes. Therefore the fact that *M.tuberculosis* apparently contains functionally active protein tyrosine kinase is very interesting and important for understanding the mechanisms of microorganism-host cell interaction, and indicates that mycobacteria apparently possess very powerful tools to control eukaryotic cell metabolism overall, and bactericidal reactions in particular.

OTHER POTENTIAL REGULATORY PRODUCTS

The fact that different mycobacterial components are able to modulate eukaryotic cell functions is well documented. The problem is that in most investigations crude preparations of mycobacterial products have been used. For example, it has been shown that glycopeptidolipids and sulfatides conferred immunosuppressive activity toward T cell functions.[54,290] In other studies polysaccharides from *M.tuberculosis* stimulated prostaglandin production of macrophages thus inhibiting several T cell functions.[207] In a recent study crude lipids of *M.intracellulare* have been shown to inhibit concanavalin A-induced T cell blastogenesis.[379] During experiments on fractionation of this crude preparation it was observed that the phospholipid fraction possessed the highest inhibitory activity. In separate experiments authors demonstrated that the observed inhibition had been caused largely by reducing interleukin-2 production and by downregulation of IL-2 receptors in ConA-stimulated T cells. Bearing in mind the very diverse spectrum of activities of lipids and in particular of phospholipids during signaling pathways in eukaryotic cells, it is not inexplicable that mycobacterial cell wall components are able to influence various reactions of macrophages, T cells and other eukaryotic cells.

In addition to traditional biochemical approaches used to investigate biological activities of mycobacterial components, molecular genetic methodology could be very fruitful in determination of novel virulence factors in *Mycobacterum*. One of these ways includes identification of genes which are specifically expressed in the virulent strain, during intracellular multiplication of mycobacteria or under stress conditions.

For example, a technique for determination of differentially expressed genes of virulent *M.tuberculosis* was developed.[203] To identify these genes authors used two strains derived from the classic H37 strain of *M.tuberculosis*—virulent H37Rv and avirulent H37Ra variants. Total RNA was isolated from both strains and RNA from H37Rv was subjected to subtractive hybridization by RNA from H37Ra. After that, a gene bank of *M.tuberculosis* was screened by a subtracted probe and eight recombinants were isolated. The detailed characterization of these cloned fragments would allow identification of new products apparently participated in a virulence of *M.tuberculosis*.

In another study[303] a similar subtractive approach has been used to identify gene(s) specifically induced in *M.avium* during their growth in macrophages. Mycobacteria grown in broth cultures were used as a source of subtractive probe whereas intracellularly replicating microorganisms supplied the target RNA. Following subtractive and hybridization procedures one DNA fragment was identified which was highly specific for *M.avium* grown in phagosomes. As for the previous investigation, the characterization of cloned gene(s) could be very promising in elucidation of strategies utilized by mycobacteria to avoid bactericidal mechanisms of phagocytes and to successfully replicate inside phagocytic cells.

In a study by Lee and Horwitz[221] authors investigated protein profiles of *M.tuberculosis* grown under stress conditions—intracellular growth, heat-shock, low pH-shock and oxygen-shock. The protein composition of the mycobacteria was unique in each case. However, of 16 proteins expressed by intracellular *M.tuberculosis*, 6 components were absent in microorganisms grown under various stress conditions and may potentially represent virulence factors specifically induced inside phagocytes.

In addition to various lipid and polysaccharide fractions possessing biological activities a number of proteins have been identified and characterized in *Mycobacterium*. Among them were a 65 kDa stress protein,[4] a 38 kDa immunodominant antigen,[78] a 6 kDa T cell antigen[350] and some others. However their role if any in intracellular multiplication of mycobacteria awaits further study.

SECTION III

CONCLUSION

The mechanisms of intracellular parasitism of pathogenic bacteria present certainly one of the most complicated and interesting problems in medical microbiology. Impressive achievements in understanding the complex character of host cell-parasite interaction have been made. However the whole problem remains far from resolution. In particular, the biology of intracellular parasitism of *Legionella*, *Listeria* and *Mycobacterium* is well studied at the morphological level. However the knowledge of bacterial products contributing to intracellular replication of these pathogens is in many instances lacking.

Apparently, to shed new light on the problem of intracellular parasitism of microorganisms new approaches are mandatory. One of them includes analysis of the interaction of phagocyte and microorganism from the point of view of signaling processes in eukaryotic cells. Such an approach has been used in this book.

The idea is that one of the modes by which intracellularly multiplying bacteria could change their host's metabolism and thus influence the normal response of phagocytes to invasive microorganisms is by affecting signaling cascades in eukaryotes. This goal could be achieved by microbes through two principal ways: (1) By binding of the bacterium to a certain receptor on the surface of a eukaryotic cell which will not initiate full-featured antibacterial attack or will guide the attached microorganism to a specific intracellular niche unreachable by the bactericidal products of phagocytes. (2) By production of enzymes and other biologically active compounds which will either mimic the action of their eukaryotic counterparts or modify the activity of key components in signaling machinery and by these means will misdirect signaling pathways for the benefit of the parasite (Fig. 22).

As outlined in the corresponding sections of this book, the number of different receptors is very broad and is paralleled by diverse spectrum of ligands available. Various types of receptors are able to initiate specific signaling pathways culminating in specific responses. Therefore, pathogenic bacteria are allowed to select the initiation of the most profitable pathway by binding a certain kind of receptor. It should be stressed that a single microorganism could engage several types of receptor molecules simultaneously. Accordingly, different cellular responses could be initiated at the same time (e.g., both stimulation of internalization of parasite and inhibition of respiratory burst in phagocyte). In this connection, investigations on mycobacteria are most remarkable. Binding of mycobacterial lipoarabinomannan to mannose receptors could be an example of initiation of a signaling cascade leading to insufficient oxidative burst and probably to other defects in the antibacterial mechanisms of phagocytes. In contrast, interaction of *Mycobacterium* with complement C3 receptors might trigger engulfment of microorganisms. These responses could be superimposed and might result in insufficient antibacterial attack toward phagocytosed parasites. Other microorganisms, namely *Listeria* and *Legionella*, also bind complement C3 receptor. However to avoid killing and degradation, listerial cells have developed their own strategy and are able to lyse phagosomes and replicate free in the cytoplasm. In the case of *Legionella* the situation is not so apparent and one may propose the existence of additional receptors on the surface of eukaryotic cells which could be utilized by this microbe to decrease antibacterial attack.

Once internalized, bacteria are faced with a problem to further manipulate eukaryotic metabolism and to destroy the regulation of antibacterial mechanisms in eukaryotic cells. The analysis of products synthesized by *Legionella, Listeria* and *Mycobacterium* demonstrates the existence of biologically active compounds with obvious regulatory potential in the sense that they are able to affect signal transduction in target cells. These substances are phospholipase C, protein kinases, phosphatases, peptidyl prolyl *cis/trans*

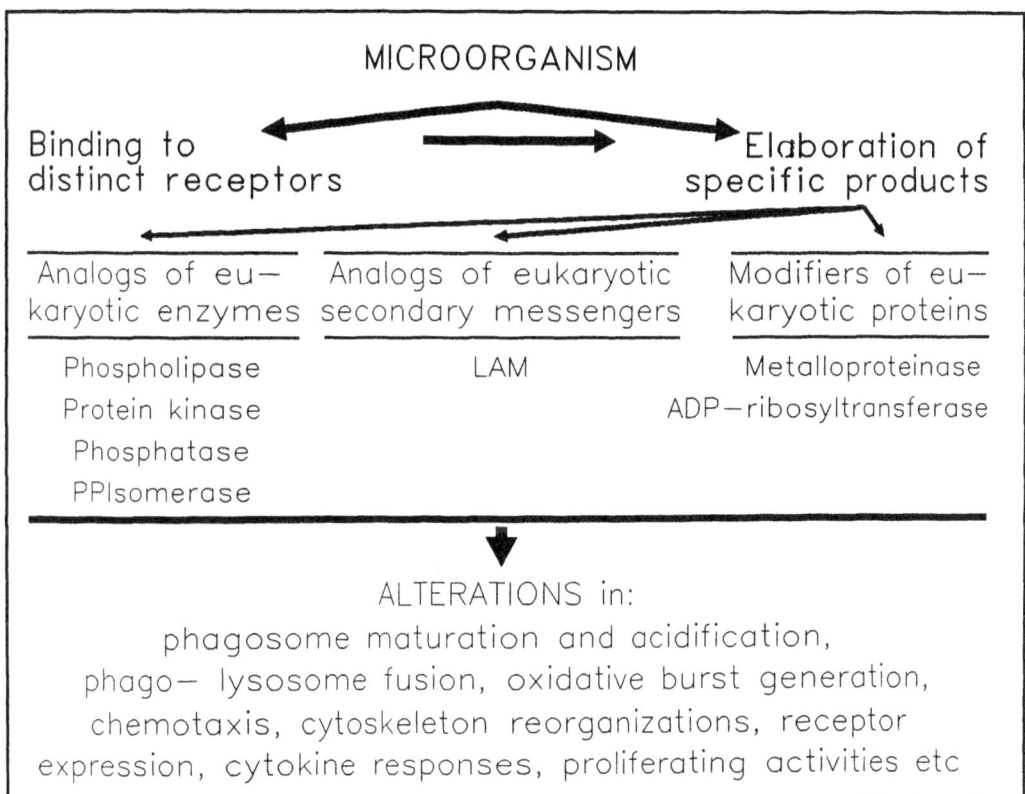

Fig. 22. Possible modes of microbial manipulations of eukaryotic signaling processes.

isomerase, ADP-ribosyltransferase, low molecular weight cytotoxin and metalloproteinase of *Legionella*, lipoarabinomannan and protein kinase of *Mycobacterium* and two phospholipases C and listeriolysin O of *Listeria*.

One prominent feature which may attract attention is disproportionate occurrence of virulence factors among these microorganisms. *Legionella* spp. produce a vast array of potential virulence factors in contrast to a rather short list of signaling products in *Listeria* and *Mycobacterium*. The reason for the more diverse spectrum of biologically active substances in *Legionella* could be the more complicated intracellular biology of this organism compared to *Listeria*. The intracellular life cycle of *Mycobacterium* in many details appears to be similar to that of *Legionella*. Therefore one may expect an equally broad range of signal-interfering products. However methodical difficulties facing a researcher (e.g., slow replication rate of mycobacteria, absence of reliable techniques for gene manipulations etc.) may explain the fact that only a few regulatory components have been identified and characterized to date.

The signal-interfering products elaborated by *Legionella*, *Listeria* and *Mycobacterium* could be divided into several groups based on the effect they accomplish. The first group might include microbial enzymes which have strong structural and functional analogies to eukaryotic proteins participating in intracellular signaling. These are phospholipases C of *Legionella* and *Listeria*, protein kinases of *Legionella* and *Mycobacterium*, phosphatases and peptidyl prolyl *cis/trans* isomerase of *Legionella*. Elaboration of such enzymes during intracellular replication of *Legionella*, *Listeria* and *Mycobacterium* could severely affect regulatory mechanisms by simulating the effect of eukaryotic enzymes in cells and thus contributing in a specific way to the survival of microbes.

It is known that besides these enzymes, other eukaryotic products participate in signal transduction in a cell as well. These are mainly ion channels, G proteins, adenylyl and guanylyl cyclases and nitric oxide synthases. In addition to these principal components there is a set of accessory proteins known to regulate the activities of the chief enzymes and ion channels or to manage the general flow of signaling cascades in eukaryotic cells. Examples of such accompanying components may include GTPase activating proteins, guanine nucleotide dissociation inhibitors and guanine nucleotide dissociation stimulators of Ras which stimulate or inhibit signal-transducing operation of these GTP-binding proteins; phosphodiesterases which destroy cyclic nucleotides thus terminating cascades of reactions accomplished through cAMP or cGMP; enzymes participating in lipidation etc. Therefore the search by biochemical or genetic methods for similar products in the above microorganisms might result in identification of novel signal-interfering proteins. This will obviously help in the elucidation of molecular mechanisms of intracellular parasitism of *Legionella*, *Listeria* and *Mycobacterium*.

The data on eukaryotic signal transduction pathways clearly indicate that in many instances the regulation of activities in key enzymes is accomplished by coordinated translocation of proteins through specialized motifs—SH2, SH3, plextrin homology domains etc. One could hypothesize that similar mechanisms may be used by prokaryotic organisms. Therefore search for these domains, or those closely related to SH2, SH3 or plextrin homology motifs, might result in detection of new regulatory microbial enzymes able to be specifically translocated to the site of their action. It is possible also that bacterial proteins, composed by analogy to Nck or Grb2 solely of SH2 or/and SH3 sequences, could be identified in the future.

The second group of potential virulence factors may include non-enzymatic components of the microbes. Such components could apparently mimic the action of second messengers in eukaryotic signal transduction and could therefore be termed "false second messengers". The mechanism of action of such bacterial products is linked to inhibition of certain signaling pathways by competition with eukaryotic second messengers for natural targets. One example of such false second messengers of bacterial origin is probably lipoarabinomannan secreted by *Mycobacterium*. It is tempting to speculate that this glycosylphospholipid with its unusual composition of fatty acid chains could be hydrolyzed by some phospholipase C to produce diacylglycerol. After binding to protein kinase C, this form of DAG could apparently compete with diacylglycerol of eukaryotic origin thus inhibiting protein kinase activity.

The third group of regulatory products of microorganisms could be enzymes which are able to modify the components of eukaryotic signaling pathways. The representatives of such enzymes are metalloproteinase and ADP-ribosyltransferase of *Legionella*. The former enzyme has been shown to specifically split components of the protein kinase system in eukaryotic cells. In relation to the infectious agent of listeriosis it is not clear if metalloproteinase of *Listeria* has any direct effect on signaling events in eukaryotic cells.

ADP-ribosyltransferase activity is one of the powerful mechanisms possessed by bacteria to inactivate certain components of signaling machinery.[21] The well known examples are pertussis toxin, LT and LT-like toxins of *Escherichia coli*, cholera and cholera-like toxins, which modify eukaryotic G proteins by covalent attachment of an ADP-ribose moiety to certain amino acid residues. Similarly, ADP-ribosyltransferase of *Legionella* has been demonstrated to modify cellular Ras-related GTP-binding proteins. It could be rewarding to search for ADP-ribosylating activity among other intracellular parasites also. In this connection it is very interesting that the existence of gene(s) related to cholera toxin has been documented in *Listeria*.

Two products of *Listeria* and *Legionella*, listeriolysin O and low molecular weight toxin correspondingly, cannot be included into any of the described groups easily. They have distinct effects on signaling processes in eukaryotic cells, which however are accomplished by unclear mechanisms. The data on these proteins suggest that the mode of their action is non-enzymatic. More detailed investigations into these bacterial products will help elucidate which steps in signaling pathways are damaged, and by which means.

The signaling pathways in eukaryotic cells have been attracting intense attention from specialists studying metabolism, transformation, differentiation, regulation of functions and other fundamental processes in eukaryotic cells. Recent data demonstrate however that the problem of signal transduction is even much broader than had been estimated earlier and spreads also into the field of interest of microbiologists studying mechanisms of intracellular parasitism of pathogenic bacteria. It turns out that intracellular parasites have the means to communicate with intracellular signaling machinery by secreting specific products. This overlay of virulence mechanisms of microorganisms and mechanisms of signal transduction in eukaryotes has resulted in a view of the problem of intracellular parasitism of bacteria. It implies that malfunctions in eukaryotic intracellular signaling, achieved through the action of specialized prokaryotic products, may represent a considerable advantage for pathogens and permit them to freely proliferate inside host cells. The latter hypothesis could form the background for future research in the area of bacterial virulence and pathogenesis of infectious diseases.

REFERENCES

1. Abramson SB, Leszczynska-Piziak J, Weissmann G. Arachidonic acid as a second messenger. Interaction with a GTP-binding protein of human neutrophils. J Immunol 1991; 147:231-236.
2. Abu-Kwaik Y, Eisenstein BI, Engleberg NC. Phenotypic modulation by *Legionella pneumophila* upon infection of macrophages. Infect Immun 1993; 61:1320-1329.
3. Adari H, Lowy DR, Willumsen BM, Der CJ, McCormick F. GTPase activating protein (GAP) interacts with the p21 Ras effector binding domain. Science 1988; 240:518-521.
4. Adebajo AO, Williams DG, Hazleman BL, Maini RN. Antibodies to the 65 kDa mycobacterial stress protein in west Africans with rheumatoid arthritis, tuberculosis and malaria. Br J Rheumatol 1995; 34:352-354.
5. Ahn NG. The MAP kinase cascade. Discovery of a new signal transduction pathway. Mol Cell Biochem 1993; 127/128:210-209.
6. Aitken A, Bilham T, Cohen P. Complete primary structure of protein phosphatase inhibitor-1 from rabbit skeletal muscle. Eur J Biochem 1992; 126:253-246.
7. Anderson D, Koch CA, Grey L, Ellis C, Moran MF, Pawson T. Binding of SH2 domains of phospholipase Cγ1, GAP and Src to activated growth factor receptor. Science 1990; 250:979-982.
8. Anderson NG, Maller JL, Tonks NK, Sturgill TW. Requirement for integration of signals from two distinct phosphorylation pathways for activation of MAP kinase. Nature 1990; 343:651-653.
9. Anderson RGW, Kamen BA, Rothberg KG, Lacey SW. Potocytosis: sequestration and transport of small molecules by caveolae. Science 1992; 255:410-411.
10. Araki S, Kikuchi K, Hata Y, Isomura M, Takai Y. Regulation of reversible binding of smg p25A, a Ras p21-like GTP-binding protein, to synaptic plasma membranes and vesicles by its specific regulatory protein, GDP dissociation inhibitor. J Biol Chem 1990; 265:13007-13015.
11. Armstrong JA, Hart PD. Response of cultured macrophages to *M.tuberculosis* with observations on fusion of lysosomes with phagosomes. J Exp Med 1971; 134:713-740.
12. Bagrodia S, Taylor SJ, Shalloway D. Myristoylation is required for Tyr-527 dephosphorylation and activation of pp60c-src in mitosis. Mol Cell Biol 1993; 13:1464-1470.

13. Baine WB. Cytolytic and phospholipase C activity in *Legionella* species. J Gen Microbiol 1985; 131:1383-1391.

14. Baine WB. A phospholipase C from the Dallas 1E strain of *Legionella pneumophila* serogroup 5: purification and characterization of conditions for optimal activity with an artificial substrate. J Gen Microbiol 1988; 134:489-498.

15. Baine WB, Rasheed JK, Mackel DC, Bopp CA, Wells JG, Kaufmann AF. Exotoxin activity associated with the Legionnaires' disease bacterium. J Clin Microbiol 1979; 9:453-456.

16. Bamezai A, Goldmacher V, Reiser H, Rock KL. Internalization of phosphatidylinositol-anchored lymphocyte proteins. I. Documentation and potential significance for T cell stimulation. J Immunol 1989; 143:3107-3116.

17. Barbacid M. *Ras* genes. Annu Rev Biochem 1987; 56:779-827.

18. Bauldry SA, Bass DA, Cousart SL, McCall SE. Tumor necrosis factor-α priming of phospholipase D in human neutrophils: correlation between phosphatidic acid production and superoxide generation. J Biol Chem 1991; 266:4173-4179.

19. Bellinger-Kawahara C, Horwitz MA. Complement component C3 fixes selectively to the major outer membrane protein (MOMP) of *Legionella pneumophila* and mediates phagocytosis of liposome-MOMP complexes by human monocytes. J Exp Med 1990; 172:1201-1210.

20. Belyi YF. Action of *Legionella* cytolysin on components of the phosphokinase system of eukaryotic cells. Biomed Sci 1990; 1:494-498.

21. Belyi YF. Intracellular parasitism of *Legionella* and signaling in eukaryotic cells. FASEB J 1993; 7:1011-1015.

22. Belyi YF, Tartakovskii IS, Vertiev YV, Prosorovskii SV. ADP-ribosyltransferase activity of *Legionella pneumophila* is stimulated by the presence of macrophage lysates. Biomed Sci 1991; 2:94-96.

23. Belyi YF, Tartakovskii IS, Vertiev YV, Prosorovskii SV. Partial purification and characterization of ADP-ribosyltransferase produced by *Legionella pneumophila*. Biomed Sci 1991; 2:169-174.

24. Belyi YF, Vertiev YV, Tartakovskii IS, Ezepchuk YV. Partial purification and characterization of a thermolabile cytotoxin produced by *Legionella pneumophila*. In: Book of Abstracts. The 1st National Conference on Molecular Structure of Bacterial Toxins and Genetic Control of their Biosynthesis. Moscow, Russia 1985:1.

25. Berger KH, Isberg RR. Two distinct defects in intracellular growth complemented by a single genetic locus in *Legionella pneumophila*. Mol Microbiol 1993; 7:7-19.

26. Berger KH, Merriam JJ, Isberg RR. Altered intracellular targeting properties associated with mutations in the *Legionella pneumophila* *dotA* gene. Mol Microbiol 1994; 14:809-822.

27. Bergman M, Mustelin T, Oetken C, Partanen J, Flint NA, Amrein KE, Autero M, Burn P, Alitalo KK. The human p50csk tyrosine kinase phosphorylates p56lck at tyr-505 and down regulates its catalytic activity. EMBO J 1992; 11:2919-2924.

28. Bermudez LE, Young LS, Enkel H. Interaction of *Mycobacterium avium* complex with human macrophages: roles of membrane receptors and serum proteins. Infect Immun 1991; 59:1697-1702.

29. Berridge MJ. Inositol triphosphate and calcium signalling. Nature 1993; 361:315-325.

30. Berridge MJ, Irvine RF. Inositol phosphates and cell signalling. Nature 1989; 341:197-205.

31. Beullens M, van Eynde A, Stalmans W, Bollen M. The isolation of novel inhibitory polypeptides of protein phosphatase 1 from bovine thymus nuclei. J Biol Chem 1992; 267:16538-16544.

32. Bhakdi S, Tranum-Jensen J, Sziegoleit A. Mechanism of membrane damage by streptolysin O. Infect Immun 1985; 47:52-60.

33. Bielecki J, Youngman P, Connelly P, Portnoy DA. *Bacillus subtilis* expressing a haemolysin gene from *Listeria monocytogenes* can grow in mammalian cells. Nature 1990; 345:175-176.

34. Billah MM. Phospholipase D and cell signaling. Curr Opin Immunol 1993; 5:114-123.

35. Birnbaumer L. G proteins in signal transduction. Annu Rev Pharmacol Toxicol 1990; 30:675-705.

36. Birnbaumer L, Abramowitz J, Brown AM. Receptor-effector coupling by G proteins. Biochim Biophys Acta Rev Biomembr 1990; 1031:163-224.

37. Black WJ, Quinn FD, Tompkins LS. *Legionella pneumophila* zinc metalloprotease is structurally and functionally homologous to *Pseudomonas aeruginosa* elastase. J Bacteriol 1990; 172:2608-2613.

38. Blander SJ, Horwitz MA. Vaccination with the major secretory protein of *Legionella pneumophila* induces cell-mediated and protective immunity in a guinea pig model of Legionnaires' disease. J Exp Med 1989; 169:691-705.

39. Blander SJ, Szeto L, Shuman HA, Horwitz MA. An immunoprotective molecule, the major secretory protein of *Legionella pneumophila* is not a virulence factor in a guinea pig model of Legionnaires' disease. J Clin Invest 1990; 86:817-824.

40. Bocckino SB, Blackmore PF, Exton JH. Stimulation of 1,2-diacylglycerol accumulation in hepatocytes by vasopressin, epinephrine and angiotensine II. J Biol Chem 1985; 260:14201-14207.

41. Bohne J, Socolovic Z, Goebel W. Transcriptional regulation of PrfA and PrfA-regulated virulence genes in *Listeria monocytogenes*. Mol Microbiol 1994; 11:1141-1150.

42. Bokoch GM. Regulation of the phagocyte respiratory burst by small GTP-binding proteins. Trends Cell Biol 1995; 5:109-113.

43. Bolen JB, Rowley RB, Spana C, Tsygankov AY. The Src family of tyrosine protein kinases in hemopoietic signal transduction. FASEB J 1992; 6:3403-3409.

44. Bollag G, McCormick F. Regulators and effectors of Ras proteins. Annu Rev Cell Biol 1991; 7:601-633.

45. Bollag G, McCormick F. GTPase activating proteins. Cancer Biol 1992; 3:199-208.

46. Bollen M, Stalmans W. The structure, role, and regulation of type 1 protein phosphatases. CRC Crit Rev Biochem Mol Biol 1992; 27:227-281.

47. Bonfini L, Karlovich CA, Dasgupta C, Banerjee U. The son of sevenless gene product: a putative activator of Ras. Science 1992; 255:603-606.

48. Brand BC, Sadosky AB, Shuman HA. The *Legionella pneumophila icm* locus: a set of genes required for intracellular multiplication in human macrophages. Mol Microbiol 1994; 14:797-808.

49. Brautigan DL, Sunwoo J, Labbe J-C, Fernandez A, Lamb NJC. Cell cycle oscillation of phosphatase inhibitor-2 in rat fibroblasts coincident with $p34^{cdc2}$ restriction. Nature 1990; 344:74-78.

50. Brennan PJ, Nikaido H. The envelope of Mycobacteria. Annu Rev Biochem 1995; 64:29-63.

51. Brodbeck U, Butikofer P. GPI anchor-hydrolyzing phospholipases. Brazilian J Med Biol Res 1994; 27:369-374.

52. Broek D, Toda T, Michaeli T, Levin L, Birchmeier C, Zoller M, Powers S, Wigler M. The *S.cerevisiae* CDC25 gene product regulates the Ras/adenylate cyclase pathway. Cell 1987; 48:789-799.

53. Brown D. The tyrosine kinase connection: how GPI-anchored proteins activate T cells. Curr Biol 1993; 5:349-354.

54. Brownback PE, Barrow WW. Modified lymphocyte response to mitogens after intraperitoneal injection of glycopeptidolipid antigens from *Mycobacterium avium* complex. Infect Immun 1988; 56:1044-1050.

55. Brundage RA, Smith GA, Camilli A, Theriot JA, Portnoy DA. Expression and phosphorylation of the *Listeria monocytogenes* ActA protein in mammalian cells. Proc Natl Acad Sci USA 1993; 90:11890-11894.

56. Brune B, Schmidt K-U, Ullrich V. Activation of soluble guanylate cyclase by carbon monoxide and inhibition by superoxide anion. Eur J Biochem 1990; 192:683-688.

57. Buechler WA, Nakane M, Murad F. Expression of soluble guanylate cyclase activity requires both enzyme subunits. Biochem Biophys Res Commun 1991; 174:351-357.

58. Burgess KE, Yamamoto M, Prasad KVS, Rudd CE. CD5 acts as a tyrosine kinase substrate within a receptor complex comprising T cell receptor zeta chain/CD3 and protein-tyrosine kinases p56[lck] and p59[fyn]. Proc Natl Acad Sci USA 1992; 89:9311-9315.

59. Butler CA, Street ED, Hatch TP, Hoffman PS. Disulfide-bonded outer membrane proteins in the genus *Legionella*. Infect Immun 1985; 48:14-18.

60. Camilli A, Goldfine H, Portnoy DA. *Listeria monocytogenes* mutants lacking phosphatidylinositol-specific phospholipase C are avirulent. J Exp Med 1991; 173:751-754.

61. Camilli A, Tilney LG, Portnoy DA. Dual roles of *plcA* in *Listeria monocytogenes* pathogenesis. Mol Microbiol 1993; 8:143-157.

62. Casabiell X, Pandiella A, Casanueva FF. Regulation of epidermal-growth-factor-receptor signal transduction by *cis*-unsaturated fatty acids. Biochem J 1991; 278:679-687.

63. Casey PJ. Lipid modifications of G proteins. Curr Op Cell Biol 1994; 6:219-225.

64. Casey PJ. Protein lipidation in cell signalling. Science 1995; 268:221-225.

65. Chakraborty T, Domann E, Ebel F, Hain T, Pistor S, Gerstel B, Niebuhr K, Lingnau A, Wehland J. Invasion and intracellular motility: two fundamental properties of pathogenic *Listeria*. In: Book of Abstracts. ISOPOL XII, Perth, Australia. 1995:495-499.

66. Chalifour R, Kanfer JN. Fatty acid activation and temperature perturbation of rat brain microsomal phospholipase D. Neurochem 1982; 39:299-305.

67. Chan AC, Desai DM, Weiss A. The role of protein tyrosine kinases and protein tyrosine phosphatases in T cell antigen receptor signal transduction. Annu Rev Immunol 1994; 12:555-592.

68. Chan AC, Iwashima M, Turck CW, Weiss A. ZAP-70 kd protein-tyrosine kinase that associates with the TCR-zeta chain. Cell 1992; 71:649-662.

69. Chan J, Fan X, Hunter SW, Brennan PJ, Bloom BR. Lipoarabinomannan, a possible virulence factor involved in persistence of *Mycobacterium tuberculosis* within macrophages. Infect Immun 1991; 59:1755-1761.

70. Chang J-H, Wilson LK, Moyers JS, Zhang K, Parsons SJ. Increased levels of p21[ras]-GTP and enhanced DNA synthesis accompany elevated tyrosyl phosphorylation of GAP-associated proteins, p190 and p62, in c-*src* overexpressors. Oncogene 1993; 8:959-967.

71. Charbonneau H, Tonks NK. 1002 protein phosphatases? Annu Rev Cell Biol 1992; 8:463-493.

72. Chen J, Martin BL, Brautigan DL. Regulation of protein serine-threonine phosphatase type 2A by tyrosine phosphorylation. Science 1992; 257:1261-1264.

73. Chernoff J, Li H-C, Cheng Y-SE, Chen LB. Characterization of a phosphotyrosyl protein phosphatase activity associated with a phosphoseryl protein phosphatase of M_r = 95,000 from bovine heart. J Biol Chem 1983; 258:7852-7857.

74. Chicurel M, Garcia E, Goodsaid F. Modulation of macrophage lysosomal pH by *Mycobacterium tuberculosis*-derived proteins. Infect Immun 1988; 56:479-483.

75. Chinkers M, Garbers DL. The protein kinase domain of the ANP receptor is required for signaling. Science 1989; 245:1392-1394.

76. Chinkers M, Garbers DL. Signal transduction by guanylyl cyclases. Annu Rev Biochem 1991; 60:553-575.

77. Chinkers M, Wilson EM. Ligand-independent oligomerization of natriuretic peptide receptors. Identification of heteromeric receptors and a dominant negative mutant. J Biol Chem 1992; 267:18589-18597.

78. Choudhary A, Vyas MN, Vyas NK, Chang Z, Quiocho FA. Crystallization and preliminary X-ray crystallographic analysis of the 38-kDa immunodominant antigen of *Mycobacterium tuberculosis*. Protein Sci 1994; 3:2450-2451.

79. Chow K, Ng D, Stokes R, Johnson P. Protein tyrosine phosphorylation in *Mycobacterium tuberculosis*. FEMS Microbiol Lett 1994; 124:203-208.

80. Cianciotto NP, Eisenstein BI, Mody CH, Engleberg NC. A mutation of *mip* gene results in an attenuation of *Legionella pneumophila* virulence. J Infect Dis 1990; 162:121-126.

81. Cianciotto NP, Eisenstein BI, Mody CH, Toews GB, Engleberg NC. A *Legionella pneumophila* gene encoding a species-specific surface protein potentiates initiation of intracellular infection. Infect Immun 1989; 57:1255-1262.

82. Cianciotto NP, Fields BS. *Legionella pneumophila mip* gene potentiates intracellular infection of protozoa and human macrophages. Proc Natl Acad Sci USA 1992; 89:5188-5191.

83. Cianciotto NP, Long R, Eisenstein BI, Engleberg NC. Site-specific mutagenesis in *Legionella pneumophila* by allelic exchange using counterselectable ColE1 vectors. FEMS Microbiol Lett 1988; 56:203-208.

84. Cianciotto NP, Stamos JK, Kamp DW. Infectivity of *Legionella pneumophila mip* mutant for alveolar epithelial cells. Curr Microbiol 1995; 30:247-250.

85. Clemens DL, Horwitz MA. Membrane sorting during phagocytosis: selective exclusion of major histocompatibility complex molecules but not complement receptor CR3 during conventional and coiling phagocytosis. J Exp Med 1992; 175:1317-1326.

86. Clemens DL, Horwitz MA. Hypoexpression of major histocompat-

ibility complex molecules on *Legionella pneumophila* phagosomes and phagolysosomes. Infect Immun 1993; 61:2803-2812.

87. Clemens DL, Horwitz MA. Characterization of the *Mycobacterium tuberculosis* phagosome and evidence that phagosomal maturation is inhibited. J Exp Med 1995; 181:257-270.

88. Colbran JL, Francis SH, Leach AB, Thomas MK, Jiang H, McAllister LM, Corbin JD. A phenylalanine in peptide substrates provides for selectivity between cGMP-dependent and cAMP-dependent protein kinases. J Biol Chem 1992; 267:9589-9594.

89. Conlan JW, Baskerville A, Ashwort LAE. Separation of *Legionella pneumophila* proteases and purification of a protease which produces lesions like those of Legionnaires' disease in guinea pig lungs. J Gen Microbiol 1986; 132:1565-1574.

90. Conlan JW, North R. Neutrophil-mediated dissolution of infected host cells as a defense strategy against a facultative intracellular bacterium. J Exp Med 1991; 174:741-744.

91. Cook SJ, Wakelam MJ. Epidermal growth factor increases sn-1,2-diacylglycerol levels and activates phospholipase D-catalyzed phosphatidylcholine breakdown in Swiss 3T3 cells in the absence of inositol-lipid hydrolysis. Biochem J 1992; 285:247-253.

92. Cooper DMF, Mons N, Fagan K. Ca^{2+}-sensitive adenylyl cyclases. Cell Signal 1994; 6:823-840.

93. Cooper DMF, Mons N, Karpen JW. Adenylyl cyclases and the interaction between calcium and cAMP signalling. Nature 1995; 374:421-424.

94. Corbin JD, Thomas MK, Wolfe L, Shabb JB, Woodford TA, Francis SH. New insights into cGMP action. In: Nishizuka Y, ed. The Biology and Medicine of Signal Transduction. New York: Raven Press, 1990:411-418.

95. Cossart P, Vincente MF, Mengaud J, Baquero F, Perez-Diaz JC, Berche P. Listeriolysin O is essential for virulence of *Listeria monocytogenes*: direct evidence obtained by gene complementation. Infect Immun 1989; 57:3929-3936.

96. Davies AA, Ley SC, Crumpton MJ. CD5 is phosphorylated on tyrosine after stimulation of the T cell antigen receptor complex. Proc Natl Acad Sci USA 1992; 89:6368-6372.

97. Dekker LV, Parker PJ. Protein kinase C—a question of specificity. Trends Biochem Sci 1994; 19:73-77.

98. Demuth A, Chakraborty T, Krohne G, Goebel W. Mammalian cells transfected with the listeriolysin gene exhibit enhanced proliferation and focus formation. Infect Immun 1994; 62:5102-5111.

99. Dent P, Lavoinne A, Nakielny S, Caudwell FB, Watt P, Cohen P. The molecular mechanism by which insulin stimulates glycogen synthesis in mammalian skeletal muscle. Nature 1990; 348:302-308.

100. Denton RM, McCormack JG. Ca^{2+} as a second messenger within mitochondria of the heart and other tissues. Annu Rev Physiol 1990; 52:451-466.

101. Domann E, Leimeister-Wachter M, Goebel W, Chakraborty T. Molecular cloning, sequencing, and identification of a metallo-proteinase gene from *Listeria monocytogenes* that is species specific and physically linked to the listeriolysin gene. Infect Immun 1991; 59:65-72.

102. Dowling JN, Saha AK, Glew RH. Virulence factors of the family *Legionellaceae*. Microbiol Rev 1982; 56:32-60.

103. Downes CP, Mussat MC, Michel RH. The inositol triphosphate phosphomonoesterase of the human erythrocyte membrane. Biochem J 1982; 203:169-177.

104. Downward J. Exchange rate mechanisms. Nature 1992; 358: 282-283.

105. Downward J, Graves JD, Warne PH, Rayter S, Cantrell DA. Stimulation of p21^{ras} upon T cell activation. Nature 1990; 346:719-723.

106. Dramsi S, Biswas I, Maguin E, Braun L, Mastroeni P, Cossart P. Entry of *Listeria monocytogenes* into hepatocytes requires expression of InlB, a surface protein of the internalin multigene family. Mol Microbiol 1995; 16:251-261.

107. Drevets D, Canono BP, Campbell PA. Listericidal and nonlistericidal mouse macrophages differ in complement receptor type 3-mediated phagocytosis of *L.monocytogenes* and in preventing escape of the bacteria into the cytoplasm. J Leukocyt Biol 1992; 52:70-79.

108. Dreyfus LA, Iglewski BH. Purification and characterization of an extracellular protease of *Legionella pneumophila*. Infect Immun 1986; 51:736-743.

109. Ebel F, Gerstel B, Pistor S, Niebuhr K, Domann E, Wehland J, Chakraborty T. A novel eukaryotic protein associated with intracytoplasmic motility of *Listeria monocytogenes*. In: Book of Abstracts. ISOPOL XII, Perth, Australia, 1995:309-313.

110. Egan SE, Giddings BW, Brooks MW, Buday L, Sizeland AM, Weinberg RA. Association of SOS Ras exchange protein with Grb2 is implicated in tyrosine kinase signal transduction and transformation. Nature 1993; 363:45-51.

111. Egerton M, Ashe OR, Chen D, Druker BJ, Burgess WH, Samelson LE. VCP, the mammalian homolog of cdc48, is tyrosine phosphorylated in response to T cell antigen receptor activation. EMBO J 1992; 11:3533-3540.

112. Egerton M, Burgess WH, Chen D, Druker BJ, Bretscher A, Samelson LE. Identification of ezrin as an 81-kDa tyrosine-phosphorylated protein in T cells. J Immunol 1992; 149:1847-1852.

113. Eichmann K. Transmembrane signaling of T lymphocytes by ligand-

induced receptor complex assembly. Angew Chem Int Ed Engl 1993; 32:54-63.

114. Engleberg NC, Carter C, Weber DR, Cianciotto NP, Eisenstein BI. DNA sequence of *mip*, a *Legionella pneumophila* gene associated with macrophage infectivity. Infect Immun 1989; 57: 1263-1270.

115. Engleberg NC, Cianciotto NP, Smith J, Eisenstein BI. Transfer and maintenance of small, mobilizable plasmids with ColE1 replication origins in *Legionella pneumophila*. Plasmid 1988; 20:83-91.

116. Fallman M, Gullberg M, Hellberg C, Andersson T. Complement receptor-mediated phagocytosis is associated with accumulation of phosphatidylcholine-derived diglyceride in human neutrophils: involvement of phospholipase D and direct evidence for a positive feedback signal of protein kinase. J Biol Chem 1992; 267: 2656-2663.

117. Fantl WJ, Escobedo JA, Martin GA, Turck CW, del Rosario M, McCormick F, Williams LT. Distinct phosphotyrosines on a growth factor receptor bind to specific molecules that mediate different signaling pathways. Cell 1992; 69:413-423.

118. Farnsworth CL, Marshall MS, Gibbs JB, Stacey DW, Feig LA. Preferential inhibition of the oncogenic form of RasH by mutations in the GAP binding/"effector" domain. Cell 1991; 64:625-633.

119. Fischer EH, Charbonneau H, Tonks NK. Protein tyrosine phosphatases: a diverse family of intracellular and transmembrane enzymes. Science 1991; 253:401-406.

120. Fischer G, Bang H, Ludwig B, Mann K, Hacker J. Mip protein of *Legionella pneumophila* exhibits peptidyl-prolyl-*cis/trans* isomerase (PPIase) activity. Mol Microbiol 1992; 6:1375-1383.

121. Francis KP, Rees CED, Stewart SAB. Identification of a major cold shock protein homologue in *Listeria monocytogenes*. In: Book of Abstracts. ISOPOL XII, Perth, Australia, 1995:403-409.

122. Friedman M, Klein TW, Friedman H. *Legionella pneumophila*-induced suppression of macrophage speading in vitro. Infect Immun 1983; 42:421-423.

123. Friedman RL, Iglewski BH, Miller RD. Identification of a cytotoxin produced by *Legionella pneumophila*. Infect Immun 1980; 29:271-274.

124. Friedman RL, Lochner JE, Bigley RH, Iglewski BH. The effects of *Legionella pneumophila* toxin on oxidative processes and bacterial killing of human polymorphonuclear leukocytes. J Infect Dis 1982; 146:328-334.

125. Fukumoto Y, Kaibuchi K, Hori Y, Fujioka H, Araki S, Ueda T, Kikuchi A, Takai Y. Molecular cloning and characterization of a novel type of regulatory protein (GDI) for the Rho proteins,

Ras p21-like small GTP-binding proteins. Oncogene 1990; 5:1321-1328.

126. Fuller F, Porter JG, Arfstein AE, Miller J, Schilling JW, Scarborough RM, Lewicki JA, Schenk DB. Atrial natriuretic peptide clearance receptor. Complete sequence and functional expression of cDNA clones. J Biol Chem 1988; 263:9395-9401.

127. Furuichi T, Mikoshiba K. Inositol 1,4,5-triphosphate receptor-mediated Ca^{2+} signaling in the brain. J Neurochem 1995; 64:953-960.

128. Furuichi T, Shiota C, Mikoshiba K. Distribution of inositol 1,4,5-triphosphate receptor mRNA in mouse tissues. FEBS Lett 1990; 267:85-88.

129. Furuichi T, Yoshikawa S, Miyawaki A, Wada K, Maeda N, Mikoshiba K. Primary structure and functional expression of the inositol 1,4,5-triphosphate-binding protein P400. Nature 1989; 342:32-38.

130. Gabay JE, Blake M, Niles WD, Horwitz MA. Purification of *Legionella pneumophila* major outer membrane protein and demonstration that it is a porin. J Bacteriol 1985; 162:85-91.

131. Gaillard JL, Berche P, Frehel C, Gouin E, Cossart P. Entry of *L.monocytogenes* into cells is mediated by internalin, a repeat protein reminiscent of surface antigens from gram-positive cocci. Cell 1991; 65:1127-1141.

132. Gaillard JL, Berche P, Mounier J, Richard S, Sansonetti PJ. In vitro model of penetration and intracellular growth of *L.monocytogenes* in the human enterocyte-like line Caco-2. Infect Immun 1987; 55:2822-2829.

133. Gandhi AJ, Perussia B, Goldfine H. *Listeria monocytogenes* phosphatidylinositol (PI)-specific phospholipase C has low activity on glycosyl-PI-anchored proteins. J Bacteriol 1993; 175:8014-8017.

134. Garbers DL. Purification of soluble guanylate cyclase from rat lung. J Biol Chem 1979; 254:240-243.

135. Garbers DL. Guanylyl cyclase receptors and their endocrine, paracrine, and autocrine ligands. Cell 1992; 71:1-4.

136. Garbers DL. Guanylyl cyclase receptors and their ligands. Adv Second Messengers Phosphorylation 1993; 28:91-95.

137. Garbers DL, Koesling D, Schultz G. Guanylyl cyclase receptors. Mol Biol Cell 1994; 5:1-5.

138. Gentschev I, Sokolovic Z, Mollenkopf HJ, Hess J, Kaufmann SH, Kuhn M, Krohne GF, Goebel W. *Salmonella* strain secreting active listeriolysin changes its intracellular localization. Infect Immun 1995; 63:4202-4205.

139. Geoffroy C, Gaillard JL, Alouf JE, Berche P. Purification, characterization and toxicity of the sulfhydryl-activated hemolysin listeriolysin O from *Listeria monocytogenes*. Infect Immun 1987; 55:1641-1646.

140. Geoffroy C, Raveneau J, Beretti J-L, Lecroisey A, Vazquez-Boland J-A, Alouf JE, Berche P. Purification and characterization of an extracellular 29-kilodalton phospholipase C from *L.monocytogenes*. Infect Immun 1991; 59:2382-2388.

141. Gerzer R, Bohme E, Hofmann F, Schultz G. Soluble guanylate cyclase purified from bovine lung contains heme and copper. FEBS Lett 1981; 132:71-74.

142. Gibson FC III, Tzianabos AO, Rodgers FG. Adherence of *Legionella pneumophila* to U-937 cells, guinea-pig alveolar macrophages, and MRC-5 cells by a novel, complement-independent binding mechanism. Can J Microbiol 1994; 10:865-872.

143. Goldfine H. The functions of the phosphatidylinositol-specific phospholipase C (PI-PLC), and the broad range phospholipase C (PC-PLC) in the pathogenesis of *Listeria monocytogenes*. In: Book of Abstracts. ISOPOL XII, Perth, Australia, 1995:315-319.

144. Goldfine H, Johnston NC, Knob C. The non-specific phospholipase C of *Listeria monocytogenes*: activity on phospholipids in Triton X-100 mixed micelles and in biological membranes. J Bacteriol 1993; 175:4298-4306.

145. Goldfine H, Knob C. Purification and characterization of *Listeria monocytogenes* phosphatidylinositol-specific phospholipase C. Infect Immun 1992; 60:4059-4067.

146. Goldfine H, Knob C, Alford D, Bentz J. Membrane permeabilization by *Listeria monocytogenes* phosphatidylinositol-specific phospholipase C is independent of phospholipid hydrolysis and cooperative with listeriolysin O. Proc Natl Acad Sci USA 1995; 92:2979-2983.

147. Gordon AH, Hart PD. Stimulation or inhibition of the respiratory burst in cultured macrophages in a *Mycobacterium* model: initial stimulation is followed by inhibition after phagocytosis. Infect Immun 1994; 62:4650-4651.

148. Graber R, Sumida C, Nunez EA. Fatty acids and cell signal transduction. J Lipid Mediators Cell Signal 1994; 9:91-116.

149. Graff JM, Stumpo DJ, Blackshear PJ. Characterization of the phosphorylation sites in the chicken and bovine myristoylated alanine-rich c-kinase substrate protein, a prominent cellular substrate for protein kinase-C. J Biol Chem 1989; 264:11912-11919.

150. Gross E, Goldberg D, Levitzki A. A phosphorylation of the *S.cerevisiae* Cdc-25 in response to glucose results in its dissociation from Ras. Nature 1992; 360:762-765.

151. Guan KL, Dixon JE. Evidence for protein-tyrosine-phosphatase catalysis proceeding via a cysteine-phosphate intermediate. J Biol Chem 1991; 266:17026-17030.

152. Gudermann T, Nurnberg B, Schultz G. Receptors and G proteins

as primary components of transmebrane signal transduction. Part 1. G protein-coupled receptors: structure and function. J Mol Med 1995; 73:51-63.

153. Gulbins E, Coggeshall KM, Baier G, Katzav S, Burn P, Altman A. Tyrosine kinase-stimulated guanine nucleotide exchange activity of Vav in T cell activation. Science 1993; 260:822-825.

154. Hacker J, Fischer G. Immunophilins: structure-function relationship and possible role in microbial pathogenicity. Mol Microbiol 1993; 10:445-456.

155. Haeffner EW. Diacylglycerol: formation and function in phospholipid-mediated signal transduction. Comp Biochem Physiol 1993; 105C:337-345.

156. Hall A. A signal transduction through small GTPases—a tale of two GAPs. Cell 1992; 69:389-391.

157. Han J-W, McCormick F, Macara IG. Regulation of Ras-GAP and the neurofibromatosis-1 gene products by eicosanoids. Science 1991; 252:576-579.

158. Hannun YA, Bell RM. Regulation of protein kinase C by sphingosine and lysosphingolipids. Clin Chim Acta 1989; 185:333-345.

159. Hart PD, Young MR, Gordon AH, Sullivan KH. Inhibition of phagosome-lysosome fusion in macrophages by certain mycobacteria can be explained by inhibition of lysosomal movements observed after phagocytosis. J Exp Med 1987; 166:933-946.

160. Hashimoto Y, Soderling TR. Regulation of calcineurin by phosphorylation. Identification of the regulatory site phosphorylated by Ca^{2+}/calmodulin-dependent protein kinase II and protein kinase C. J Biol Chem 1989; 264:16524-16529.

161. Hedin KE, Duerson K, Clapham DE. Specificity of receptor-G protein interactions: searching for the structure behind the signal. Cell Signal 1993; 5:505-518.

162. Hedlund KW. *Legionella* toxin. Pharm Ther 1981; 15:123-130.

163. Heemskerk JWM, Sage SO. Calcium signaling in platelets and other cells. Platelet 1994; 5:295-316.

164. Heldin C-H. Dimerization of cell surface receptors in signal transduction. Cell 1995; 80:213-233.

165. Hemmings HC Jr, Greengard P, Tung HYL, Cohen P. DARPP-32, a dopamine-regulated neuronal phosphoprotein, is a potent inhibitor of protein phosphatase-1. Nature 1984; 310:503-505.

166. Hess J, Gentchev I, Szalay G, Ladel C, Bubert A, Goebel W, Kaufmann SHE. *Listeria monocytogenes* p60 supports host cell invasion by and in vivo survival of attenuated *Salmonella typhimurium*. Infect Immun 1995; 63:2047-2053.

167. Hindahl MS, Iglewski BH. Outer membrane proteins from *Legionella pneumophila* serogroups and other *Legionella* species. Infect Immun 1986; 51:94-101.

168. Holmes CFB, Kuret J, Chisholm AAK, Cohen P. Identification of the sites on rabbit skeletal muscle protein phosphatase inhibitor 2 phosphorylated by casein kinase-II. Biochim Biophys Acta 1986; 870:408-416.

169. Honegger AM, Schmidt A, Ulrich A, Schlessinger J. Evidence for epidermal growth factor (EGF)-induced intermolecular autophosphorylation of the EGF receptors in living cells. Mol Cell Biol 1990; 10:4035-4044.

170. Horwitz MA. Formation of a novel phagosome by the Legionnaires' disease bacterium (*Legionella pneumophila*) in human monocytes. J Exp Med 1983; 158:1319-1331.

171. Horwitz MA. The Legionnaires' disease bacterium (*Legionella pneumophila*) inhibits phagosome-lysosome fusion in human monocytes. J Exp Med 1983; 158:2108-2126.

172. Horwitz MA. Phagocytosis of the Legionnaires' disease bacterium (*Legionella pneumophila*) occurs by a novel mechanism: engulfment within pseudopod coil. Cell 1984; 36:27-33.

173. Horwitz MA. Phagocytosis and intracellular biology of *Legionella pneumophila*. In: Bacteria-Host Cell Interaction. Alan R. Liss, Inc, 1988:283-302.

174. Horwitz MA, Maxfield FR. *Legionella pneumophila* inhibits acidification of its phagosome in human monocytes. J Cell Biol 1984; 99:1936-1943.

175. House C, Kemp BE. Protein kinase C contains a pseudosubstrate prototype in its regulatory domain. Science 1987; 238:1726-1728.

176. Hsi ED, Siegel JN, Minami Y, Luong ET, Klausner RD, Samelson LE. T cell activation induces rapid tyrosine phosphorylation of a limited number of cellular substrates. J Biol Chem 1989; 264:10836-10842.

177. Huang C, Wykle RL, Daniel LW, Cabot MC. Identification of phosphatidylcholine-selective and phosphatidylinositol-selective phospholipase D in Madine-Darby kidney cells. J Biol Chem 1992; 267:16859-16865.

178. Huang JMC, Xian H, Bacaner M. Long-chain fatty acids activate calcium channels in ventricular myocytes. Proc Natl Acad Sci USA 1992; 89:6452-6456.

179. Hughes DA, Fukui Y, Yamamoto M. Homologous activators of Ras in fission and budding yeast. Nature 1990, 344:355-357.

180. Humbert P, Nitroomand F, Fischer G, Mayer B, Koesling D, Hinsch K-D, Gausepohl H, Frank R, Schultz G, Bohme E. Purification of soluble guanylyl cyclase from bovine lung by a new immunoaffinity chromatographic method. Eur J Biochem 1990; 190:273-278.

181. Hunter T. Protein kinases and phosphatases: the Yin and Yang of protein phosphorylation and signaling. Cell 1995; 80:225-236.

182. Hunter T, Lindberg RA, Middlemas DS, Tracy S, van der Geer P. Receptor protein tyrosine kinases and phosphatases. Cold Spring Harbor Symp Quant Biol 1992; LVII: 25-41.

183. Hurley TR, Hyman R, Sefton BM. Differential effects of expression of the CD45 tyrosine protein phosphatase on the tyrosine phosphorylation of the Lck/Fyn and c-Src tyrosine kinases. Mol Cell Biol 1993; 13:1651-1656.

184. Husmann LK, Johnson W. Cytotoxicity of extracellular *Legionella pneumophila.* Infect Immun 1994; 62:2111-2114.

185. Ilangumaran S, Arni S, Poincelet M, Theler JM, Brennan PJ, Nasir-ud-Din, Hoessli DC. Integration of mycobacterial lipoarabino-mannans into glycosylphosphatidylinositol-rich domains of lymphomonocytic cell plasma membranes. J Immunol 1995; 155:1334-1342.

186. Isberg RR, Rankin S, Roy CR, Swanson MS, Berger K. *Legionella pneumophila*: factors involved in the route and response to an intracellular niche. Infect Agents Dis 1994; 2:220-223.

187. Iyengar R. Molecular and functional diversity of mammalian G_s-stimulated adenylyl cyclases. FASEB J 1993; 7:768-775.

188. Jacob T, Escallier JC, Sanguedolce MV, Chicheportiche C, Bongrand P, Capo C, Mege JL. *Legionella pneumophila* inhibits superoxide generation in human monocytes via the down-modulation of alpha and beta protein kinase C isotypes. J Leuk Biol 1994; 55:310-312.

189. Johnson P, Ostergaard HL, Wasden C, Trowbridge IS. Mutational analysis of CD45: a leukocyte-specific protein tyrosine phosphatase. J Biol Chem 1992; 12:8035-8041.

190. Jonak C, Heberle-Bors E, Hirt H. MAP kinases: universal multi-purpose signaling tools. Plant Mol Biol 1994; 24:407-416.

191. Kahn RA, Der CJ, Bokoch GM. The Ras superfamily of GTP-binding proteins: guidelines on nomenclature. FASEB J 1992; 6:2512-2513.

192. Kaibuchi K, Mizuno T, Fujioka H, Yamamoto T, Kishi K, Fukumoto Y, Hori Y, Takai Y. Molecular cloning of the cDNA for stimulatory GDP/GTP exchange protein for smg p21s (Ras p21-like small GTP-binding proteins) and characterization of stimulatory GDP/GTP exchange protein. Mol Cell Biol 1991; 11:2873-2880.

193. Kaldenberg-Stasch S, Baden M, Fessler B, Jacobs KH, Wieland T. Receptor-stimulated guinine nucleoside triphosphate binding to G proteins: nucleotide exchange and β subunit-mediated phosphotransfer reactions. Eur J Biochem 1994; 221:25-33.

194. Kamps MP, Sefton BM. Neither arginine nor histidine can carry out the function of lysine-295 in the ATP-binding site of p60^{src}.

Mol Cell Biol 1986; 6:751-757.

195. Kanaho Y, Kanoh H, Saitoh K, Nozawa Y. Phospholipase D activation by platelet-activating factor, leukotriene B$_4$ and formyl-methionylleucylphenylalanine in rabbit neutrophils: phospholipase D activation is involved in enzyme release. J Immunol 1991; 146:3536-3541.

196. Kathariou S, Metz P, Hof H, Goebel W. Tn *916*-induced mutations in the hemolysin determinant affecting virulence of *Listeria monocytogenes*. J Bacteriol 1987; 169:1291-1297.

197. Kaziro Y, Itoh H, Kozasa T, Nakafuku M, Satoh T. Structure and function of signal-transducing GTP-binding proteins. Annu Rev Biochem 1991; 60:349-400.

198. Khan WA, El Touny S, Hannun YA. Arachidonic and *cis*-unsaturated fatty acids induce selective platelet substrate phosphorylation through activation of cytosolic protein kinase C. FEBS Lett 1991; 292:98-102.

199. Khosravi-Far R, Der CJ. The Ras signal transduction pathway. Cancer Metastasis Rev 1994; 13:67-89.

200. Kilgour E. A role for inositol-glycan mediators and G proteins in insulin action. Cell Signal 1993; 5:97-105.

201. Kim MJ, Rogers JE, Hurley MC, Engleberg NC. Phosphatase-negative mutants of *Legionella pneumophila* and their behavior in mammalian cell infection. Microb Pathog 1994; 17:51-62.

202. King WG, Rittenhouse SE. Inhibition of protein kinase C by staurosporine promotes accumulation of inositol triphosphate and tetrakisphosphate in human platelets exposed to thrombin. J Biol Chem 1989; 264:6070-6074.

203. Kinger AK, Tyagi JS. Identification and cloning of genes differentially expressed in the virulent strain of *Mycobacterium tuberculosis*. Gene 1993; 131:113-117.

204. Kiss Z. Differential effects of platelet-derived growth factor, serum and bombesin on phospholipase D-mediated hydrolysis of phosphatidyl-ethanolamine in NIH 3T3 fibroblasts. Biochem J 1992; 285:229-233.

205. Klee CB, Draetta GF, Hubbard MJ. Calcineurin. Adv Enzymol 1988; 61:149-200.

206. Klee CB, Krinks MH. Purification of cyclic 3',5'-nucleotide phosphodiesterase inhibitory protein by affinity chromatography on activator protein coupled to Sepharose. Biochemistry 1978; 17:120-126.

207. Kleinhenz ME, Ellner JJ, Spagnuolo PJ, Daniel TM. Suppression of lymphocytes response by tuberculous plasma and mycobacterial arabinogalactan. Monocyte dependence and indomethacin reversibility. J Clin Invest 1981; 68:153-162.

208. Koch CA, Anderson D, Moran MF, Ellis C, Pawson T. SH2 and SH3 domains: elements that control interactions of cytoplasmic signaling proteins. Science 1991; 252:668-674.

209. Kocks C, Gouin E, Tabouret M, Berche P, Ohayon H, Cossart P. *L.monocytogenes*-induced actin assembly requires the *actA* gene product, a surface protein. Cell 1992; 68:521-531.

210. Kocks C, Hellio R, Gounon P, Ohayon H, Cossart P. Polarized distribution of *Listeria monocytogenes* surface protein ActA at the site of directional actin assembly. J Cell Sci 1993; 105:699-710.

211. Koesling D, Bohme E, Schultz G. Guanylyl cyclases, a growing family of signal-transducing enzymes. FASEB J 1991; 5:2785-2791.

212. Kolch W, Heidecker G, Kochs G, Hummel R, Vahidi H, Mischak H, Finkenzeller G, Marme D, Rapp UR. PKC α activates Raf-1 by direct phosphorylation. Nature 1993; 364:249-252.

213. Kozma SC, Thomas G. Serine/threonine kinases in the propagation of the early mitogenic response. Rev Physiol Biochem Pharmacol 1992; 119:123-155.

214. Krebs EG. The enzymology of control by phosphorylation. In: Boyer PD, ed. The Enzymes. New York: Academic Press, 1986; XVII: 3-19.

215. Kuhn M, Goebel W. Identification of an extracellular protein of *Listeria monocytogenes* possibly involved in intracellular uptake by mammalian cells. Infect Immun 1989; 57:55-61.

216. Landt M, Easom RA, Colca JR, Wolf BA, Turk J, Mills LA, McDaniel ML. Parallel effects of arachidonic acid on insulin secretion, calmodulin-dependent protein kinase activity and protein kinase C activity in pancreatic islets. Cell Calcium 1992; 13:163-172.

217. Lang K, Schmid FX, Fischer G. Catalysis of protein folding by prolyl isomerase. Nature 1987; 329:268-270.

218. Lange-Carter CA, Pleiman CM, Gardner AM, Blumer KJ, Johnson GL. A divergence in the MAP kinase regulatory network defined by MEK kinase and Raf. Science 1993; 260:315-319.

219. Leahy DJ. A structural view of CD4 and CD8. FASEB J 1995; 9:17-25.

220. Ledbetter JA, Deans JP, Aruffo A, Grosmaire LS, Kanner SB, Bolen JB, Schieven GL. CD4, CD8 and the role of CD45 in T cell activation. Curr Op Immunol 1993; 5:334-340.

221. Lee BY, Horwitz MA. Identification of macrophage and stress-induced proteins of *Mycobacterium tuberculosis*. J Clin Invest 1995; 96:245-249.

222. Lefkowitz RJ, Inglese J, Koch WJ, Pitcher J, Attramadal H, Caron MG. G protein-coupled receptors: regulatory role of receptor kinases and arrestin proteins. Cold Spring Harbor Symp Quant Biol 1992; LVII: 127-133.

223. Leimeister-Wachter M, Chakraborty T. Detection of listeriolysin, the thiol-dependent hemolysin in *Listeria monocytogenes*, *Listeria ivanovii*, and *Listeria seeligeri*. Infect Immun 1990; 57:2350-2357.

224. Leimeister-Wachter M, Domann E, Chakraborty T. Detection of a gene encoding a phosphatidylinositol-specific phospholipase C that is coordinately expressed with listeriolysin in *Listeria monocytogenes*. Mol Microbiol 1991; 5:361-366.

225. Leimeister-Wachter M, Domann E, Chakraborty T. The expression of virulence genes in *Listeria monocytogenes* is thermoregulated. J Bacteriol 1992; 174:947-952.

226. Letourneur F, Klausner RD. Activation of T cells by a tyrosine kinase activation domain in the cytoplasmic tail of CD3-ε. Science. 1992; 5:79-82.

227. Lin L, Wartmann M, Lin A, Knopf JL, Seth A, Davis RJ. cPLA$_2$ is phosphorylated and activated by MAP kinase. Cell 1993; 2:269-278.

228. Lincoln TM, Corbin JD. Characterization and biological role of the cGMP-dependent protein kinase. Adv Cyclic Nucleotide Res 1983; 15:139-192.

229. Lincoln TM, Pryzwansky KB, Cornwell TL, Wyatt T, MacMillan LA. Cyclic-GMP-dependent proteinkinase in smooth muscle and neutrophils. Adv Second Messenger Phosphorylation Res 1993; 28:121-132.

230. Lingnau A, Domann E, Hudel M, Bock M, Nichterlein T, Wehland J, Chakraborty T. Expression of the *Listeria monocytogenes* EGD *inlA* and *inlB* genes, whose products mediate bacterial entry into tissue culture cell lines by PrfA-dependent and independent mechanisms. Infect Immun 1995; 63:3896-3903.

231. Liscovitch M, Cantley LC. Lipid second messengers. Cell 1994; 77:329-334.

232. Lister MD, Deems RA, Watanabe Y, Ulevitch RJ, Dennis ES. Kinetic analysis of the Ca^{2+}-dependent, membrane-bound macrophage phospholipase A$_2$ and the effect of arachidonic acid. J Biol Chem 1988; 263:7506-7513.

233. Lochner JE, Bigley RH, Iglewski BH. Defective triggering of polymorphonuclear leukocyte oxidative metabolism by *Legionella pneumophila* toxin. J Infect Dis 1985; 151:42-46.

234. Louis JC, Basset P, Revel MO, Vincendon G, Zwiller J. Opposite effects of arachidonic acid and of its hydroperoxides on brain soluble guanylate cyclase activity. Neurochem Int 1991; 18:131-135.

235. Lowe DG. Human natriuretic peptide receptor-A guanylyl cyclase is self-associated prior to hormone binding. Biochemistry 1992; 31:10421-10425.

236. Ludwig B, Rahfeld J, Schmidt B, Mann K, Wintermeyer E, Fischer

G, Hacker J. Characterization of Mip proteins of *Legionella pneumophila*. FEMS Microbiol Lett 1994; 118:21-30.

237. Lundemose AG, Kay JE, Pearce JH. *Chlamydia trachomatis* Mip-like protein has peptidyl-prolyl *cis/trans* isomerase activity that is inhibited by FK506 and rapamycin and is implicated in initiation of chlamydial infection. Mol Microbiol 1993; 7:777-783.

238. Maeda N, Niinobe M, Nakashira K, Mikoshiba K. Purification and characterization of P400 protein, a glycoprotein characteristic of Purkinje cell from mouse cerebellum. J Neurochem 1988; 51:1724-1730.

239. Maines MD. Heme oxygenase: function, multiplicity, regulatory mechanisms, and clinical applications, FASEB J 1988; 2:2557-2568.

240. Malissen B, Schmitt-Verhulst A-M. Transmembrane signalling through the T cell-receptor-CD3 complex. Curr Op Immunol 1993; 5:324-333.

241. Martegani E, Vanoni M, Zippel R, Coccetti P, Brambilla R, Farrari C, Sturani E, Alberghina L. Cloning by functional complementation of a mouse cDNA encoding a homolog of CDC25, a *Saccharomyces cerevisiae* Ras activator. EMBO J 1992; 11:2151-2157.

242. Martin GA, Yatani A, Clark R, Conroy L, Polakis P, Brown AM, McCormick F. GAP domains responsible for Ras p21-dependent inhibition of muscarinic atrial K$^+$ channel currents. Science 1992; 255:192-194.

243. Mayer BJ, Ren R, Clark KL, Baltimore D. A putative modular domain present in diverse signaling proteins. Cell 1993; 73:629-630.

244. McCormick F. How receptors turn Ras on. Nature 1993; 363:15-16.

245. McCusker KT, Braaten BA, Cho MW, Low DA. *Legionella pneumophila* inhibits protein synthesis in Chinese hamster ovary cells. Infect Immun 1991; 59:240-246.

246. McDade JE, Shepard CC, Fraser DW, Tsai TR, Redus MA, Dowdle WR and the Laboratory Investigation Team. Legionnaires' disease: isolation of a bacterium and demonstration of its role in other respiratory diseases. N Engl J Med 1977; 297:1197-1203.

247. McDonough KA, Kress Y, Bloom BR. Pathogenesis of tuberculosis: interaction of *Mycobacterium tuberculosis* with macrophages. Infect Immun 1993; 61:2763-2773.

248. McFarland EDC, Hurley TR, Pingel JT, Sefton BM, Shaw A, Thomas ML. Correlation between Src family member regulation by the protein-tyrosine phosphatase CD45 and transmembrane signaling through the T cell receptor. Proc Natl Acad Sci USA 1993; 90:1402-1406.

249. McLauchlin J, Low C. Primary cutaneous listeriosis in adults: an occupational disease of veterinarians and farmers. Vet Res 1994;

135:615-617.

250. Medema RH, de Vries-Smits AMM, van der Zon GCM, Maasen JA, Bos JL. Ras activation by insulin and epidermal growth factor through enhanced exchange of guanine nucleotides on p21^{ras}. Mol Cell Biol 1993; 13:155-162.

251. Medema RH, Bos JL. The role of p21ras in receptor tyrosine kinase signaling. Crit Rev Oncogen 1993; 4:615-661.

252. Mengaud J, Braun-Breton C, Cossart P. Identification of a phosphatidylinositol-specific phospholipase C in *Listeria monocytogenes*: a novel type of virulence factor? Mol Microbiol 1991; 5:367-372.

253. Mengaud J, Chenevert J, Geoffroy C, Gaillard J-L, Cossart P. Identification of the structural gene encoding the SH-activated hemolysin of *Listeria monocytogenes*: listeriolysin O is homologous to streptolysin O and pneumolysin. Infect Immun 1987; 55:3225-3227.

254. Mengaud J, Vecente M-F, Chenevert J, Pereira JM, Geoffroy C, Gicquel-Sanzey B, Baquero F, Perez-Diaz JC, Cossart P. Expression in *Escherichia coli* and sequence analysis of the listeriolysin O determinant of *Listeria monocytogenes*. Infect Immun 1988; 56:766-772.

255. Menniti FS, Oliver KG, Nogimori K, Obie JF, Shears SB, Putney JW Jr. Origins of myoinositol tetrakisphosphates in agonist-stimulated rat pancreatoma cells: stimulation by bombesin of myoinositol(1,3,4,5,6)pentakisphosphate breakdown to myoinositol(3,4,5,6)tetrakisphosphate. J Biol Chem 1990; 265:11167-11176.

256. Merrill AH. Ceramide: a new lipid "second messenger"? Nutr Rev 1992; 50:78-80.

257. Merril AH, Stevens VL. Modulation of protein kinase C and diverse cell functions by sphingosine—a pharmacologically interesting compound linking sphingolipids and signal transduction. Biochim Biophys Acta 1989; 1010:131-139.

258. Michikawa T, Hamanaka H, Otsu H, Yamamoto A, Miyawaki A, Furuichi T, Tashiro Y, Mikoshiba K. Transmembrane topology and sites of N-glycosylation of inositol 1,4,5-triphosphate receptor. J Biol Chem 1994; 269:9184-9189.

259. Michnick SW, Rosen MK, Wandless TJ, Karplus PA, Schreiber SL. Solution structure of FKBP, a rotamase enzyme and receptor for FK506 and rapamycin. Science 1991; 252:836-839.

260. Mignery GA, Sudhof TC. The ligand binding site and transduction mechanism in the inositol-1,4,5-triphosphate receptor. EMBO J 1990; 9:3893-3898.

261. Mignery GA, Sudhof TG, Takei K, De Camilli P. Putative receptor for inositol 1,4,5-triphosphate similar to ryanodine receptor. Nature 1989; 342:192-195.

262. Mikoshiba K, Furuichi T, Miyawaki A, Yoshikawa S, Nakagawa T, Yamada N, Hamanaka Y, Fujino I, Michikawa T, Ryo Y, Okano H, Fujii S, Nakade S. Inositol triphosphate receptor and Ca^{2+} signalling. Phil Trans R Soc Lond 1993; 340:345-349.

263. Mikoshiba K, Huchet M, Changeux J-P. Biochemical and immunological studies on the P400 protein, a protein characteristic of the Purkinje cell from mouse and rat cerebellum. Devl Neurosci 1979; 2:254-275.

264. Mintz SC, Miller RD, Gutgsell NS, Malek T. *Legionella pneumophila* protease inactivates interleukin-1 and cleaves CD4 on human T cells. Infect Immun 1993; 61:3416-3421.

265. Miyamoto H, Yoshida S-I, Taniguchi H, Qin MH, Fujio H, Mizuguchi Y. Protein profiles of *Legionella pneumophila* Philadelphia-1 grown in macrophages and characterization of a gene encoding a novel 24 kDa *Legionella* protein. Microb Pathogen 1993; 15:469-484.

266. Mizuno T, Kaibuchi K, Yamamoto T, Kawamura M, Sakoda T, Fujioka H, Matsuura Y, Takai Y. A stimulatory GDP/GTP exchange protein for smg p21 is active on the posttranslationally processed form of c-Ki-ras p21 and rhoA p21. Proc Natl Acad Sci USA 1991; 88:6442-6446.

267. Moran MF, Polakis P, McCormick F, Pawson T, Ellis C. Protein tyrosine kinases regulate the phosphorylation, protein interactions, subcellular distribution, and activity of p21ras GTPase-activating protein. Mol Cell Biol 1991; 11:1804-1812.

268. Muder RR, Yu VL, Woo AH. Mode of transmission of *Legionella pneumophila*. Arch Intern Med 1986; 146:1607-1612.

269. Muller HE. Enzymatic profile of *Legionella pneumophila*. J Clin Microbiol 1981; 13:423-426.

270. Mumby MC, Walter G. Protein serine/threonine phosphatases: structure, regulation, and functions in cell growth. Physiol Rev 1993; 73:673-699.

271. Murakami K, Chan SY, Routtenberg A. Protein kinase C activation by *cis*-fatty acid in the absence of Ca^{2+} and phospholipids. J Biol Chem 1986; 261:15424-15429.

272. Murphy MG. Membrane fatty acids, lipid peroxidation and adenylate cyclase activity in cultured neural cells. Biochem Biophys Res Commun 1985; 132:757-763.

273. Murray EGD, Webb RE, Swann MBR. A desease of rabbits characterized by a large mononuclear leukocytosis, caused by a hitherto undescribed bacillus *Bacterium monocytogenes* (n.sp.). J Pathol Bacteriol 1926; 29:407-439.

274. Mustelin T, Coggeshall KM, Altman A. Rapid activation of the T cell tyrosine protein kinase pp56lck by the CD45 phosphotyrosine

phosphatase. Proc Natl Acad Sci USA 1989; 86:6302-6306.

275. Mustelin T, Pessa-Morikawa T, Autero M, Gassmann M, Anderson LC, Gahmberg CG, Burn P. Regulation of the p59fyn protein tyrosine kinase by the CD45 phosphotyrosine phosphatase. Eur J Immunol 1992; 22:1173-1178.

276. Nairn AC. Role of Ca^{2+}/Calmodulin-dependent protein phosphorylation in signal transduction. In: Nishizuka Y, ed. The Biology and Medicine of Signal Transduction. New York: Raven Press, 1990; 202-205.

277. Nakade S, Maeda N, Mikoshiba K. Involvement of the C-terminus of the inositol 1,4,5-triphosphate receptor in Ca^{2+} release analyzed using region-specific monoclonal antibodies. Biochem J 1991; 277:125-131.

278. Nathan C, Xie Q-W. Nitric oxide synthases: roles, tolls, and controls. Cell 1994; 78:915-918.

279. Nimmo GA, Cohen P. The regulation of glycogen metabolism. Purification and characterization of protein phosphatase inhibitor-1 from rabbit skeletal muscle. Eur J Biochem 1978; 87:341-351.

280. Nishizuka Y. Intracellular signaling by hydrolysis of phospholipids and activation of protein kinase C. Science 1992; 258:607-614.

281. Nolte FS, Conlin CA. Major outer membrane protein of *Legionella pneumophila* carries a species-specific epitope. J Clin Microbiol 1986; 23:643-646.

282. Nolte FS, Hollick GE, Robertson RG. Enzymatic activities of *Legionella pneumophila* and *Legionella*-like organisms. J Clin Microbiol 1982; 15:175—177.

283. Nurnberg B, Gudermann T, Schultz G. Receptors and G proteins as primary components of transmembrane signal transduction. Part 2. G proteins: structure and function. J Mol Med 1995; 73:123-132.

284. Offermanns S, Schultz G. Complex information processing by the transmembrane signaling system involving G proteins. Naunyn-Schmiedeberg's Arch Pharmacol 1994; 350:329-338.

285. Ogita K, Koide H, Kikkawa U, Kishinoto A, Nishizuka Y. The heterogeneity of protein kinase C in signal transduction cascade. In: Nishizuka Y, ed. The Biology and Medicine of Signal Transduction. New York: Raven Press, 1990:218-223.

286. Okada M, Nada S, Yamanashi NY, Yamamoto T, Nakagawa H. Csk: a protein-tyrosine kinase involved in regulation of *src* family kinases. J Biol Chem 1991; 266:24249-24252.

287. Oliveira L, Paiva ACM, Sander C, Vriend G. A common step for signal transduction in G protein-coupled receptors. Trends Pharmacol Sci 1994; 15:170-172.

288. Ostergaard HL, Shackelford DA, Hurley TR, Johnson P, Hyman

R, Sefton BM, Trowbridge IS. Expression of CD45 alters phosphorylation of the *lck*-encoded tyrosine protein kinase in murine lymphoma T cell lines. Proc Natl Acad Sci USA 1989; 86:8959-8963.

289. Otero AS. Transphosphorylation and G protein activation. Biochem Pharmacol 1990; 9:1399-1404.

290. Pabst MJ, Gross JM, Brozna JP, Goren MB. Inhibition of macrophage priming by sulfatide from *Mycobacterium tuberculosis*. J Immunol 1988; 140:634-640.

291. Paige LA, Nadler MJS, Harrison ML, Cassady JM, Geahlen RL. Reversible palmitoylation of the protein-tyrosine kinase p56lck. J Biol Chem 1993; 268:8669-8674.

292. Panayotou G, Waterfield MD. The assembly of signalling complexes by receptor tyrosine kinases. BioEssays 1993; 15:171-177.

293. Parks RE, Agarwal RP. Nucleoside diphosphokinases. In: Boyer PD, ed. The Enzymes. New York: Academic Press, 1973; VIII: 307-333.

294. Parmentier M, Libert F, Perret J, Eggerickx D, Ledent C, Schurmans S, Raspe E, Dumont JE, Vassart G. Cloning and characterization of G protein-coupled receptors. Adv Second Messenger Phosphoprotein Res 1993; 28:11-18.

295. Pawson T, Gish G. SH2 and SH3 domains: from structure to function. Cell 1992; 71:359-362.

296. Payne NR, Horwitz MA. Phagocytosis of *Legionella pneumophila* is mediated by human monocyte complement receptors. J Exp Med 1987; 1966:1377-1389.

297. Pearlman E, Jiwa AH, Engleberg NC, Eisenstein BI. Growth of *Legionella pneumophila* in human macrophage-like (U937) cell line. Microb Pathogen 1988; 5:87-95.

298. Peitsch M, Borner C, Tschopp J. Sequence similarity of phospholipase A2 activating protein and the G protein β subunits: a new concept of effector protein activation in signal transduction? Trends Biochem Sci 1993; 18:292-293.

299. Pelech SL. Networking with protein kinases. Curr Biol 1993; 3:513-515.

300. Pelicci G, Lanfrancone L, Grignani F, McGlade J, Cavallo F, Forni G, Nicoletti I, Pawson T, Pelicci PG. A novel transforming protein (SHC) with an SH2 domain is implicated in mitogenic signal transduction. Cell 1992; 70:93-104.

301. Perry DK, Hand WL, Edmondson DE, Lambeth JD. Role of phospholipase D-derived diacylglycerol in the activation of the human neutrophil respiratory burst oxidase: inhibition by phosphatidic acid phosphohydrolase inhibitors. J Immunol 1992; 149:2749-2758.

302. Phan-Thanh L, Gormon T. Analysis of heat and cold shock proteins in *Listeria* by two-dimensional electrophoresis. Electrophoresis 1995; 16:444-450.

303. Plum G, Clark-Curtiss JE. Induction of *Mycobacteium avium* gene expression following phagocytosis by human macrophages. Infect Immun 1994; 62:476-483.

304. Portnoy DA, Jacks PS, Hinrichs D. Role of hemolysin for the intracellular growth of *L.monocytogenes*. J Exp Med 1988; 167: 1459-1471.

305. Pot DA, Dixon JE. Active site labeling of a receptor-like protein tyrosine phosphatase. J Biol Chem 1992; 267:140-143.

306. Potter LR, Garbers DL. Dephosphorylation of the guanylyl cyclase-A receptor causes desensitization. J Biol Chem 1992; 267: 14531-14534.

307. Premont RT, Jacobowitz O, Iyengar R. Lowered responsiveness of the catalyst of adenylyl cyclase to stimulation by G_s in heterologous desensitization: a role for cAMP dependent phosphorylation. Endocrinology 1992; 131:2774-2783.

308. Putney JW Jr. Excitement about calcium signaling in inexcitable cells. Science 1993; 262:676-678.

309. Putney JW jr, Bird GSJ. The inositol phosphate-calcium signaling system in nonexcitable cells. Endocrine Rev 1993; 14:610-631.

310. Racz P, Tenner K, Mero E. Experimental *Listeria* enteritis. 1. An electron microscopic study of the epithelial phase in experimental *Listeria* infection. Lab Invest 1972; 26:694-700.

311. Rajagopalan-Levasseur P, Dournon E, Vilde JL, Pocidalo JJ. Differences in the respiratory burst of human polymorphonuclear leukocytes induced by virulent and avirulent *Legionella pneumophila* serogroup 1. J Biolumin Chemilumin 1992; 7:109-116.

312. Rankin S, Isberg RR. Identification of *Legionella pneumophila* promoters regulated by the macrophage intracellular environment. Infect Agents Dis 1994; 2:269-271.

313. Raveneau J, Geoffroy C, Beretti J-L, Gaillard J-L, Alouf JE, Berche P. Reduced virulence of a *Listeria monocytogenes* phospholipase-deficient mutant obtained by transposon insertion into the zinc metalloproteinase gene. Infect Immun 1992; 60:916-921.

314. Rechnitzer C, Diamant M, Pederson BK. Inhibition of human natural killer cell activity by *Legionella pneumophila* protease. Eur J Clin Microbiol Infect Dis 1989; 8:989-992.

315. Rechnitzer C, Kharazmi A, Nielsen H. Effects of *Legionella pneumophila* sonicate on human neutrophil granulocyte and monocyte chemotaxis. Eur J Clin Invest 1986; 16:368-375.

316. Ridley AJ, Paterson HF, Johnston CL, Diekmann D, Hall A. The small GTP-binding protein Rac regulates growth factor-induced membrane ruffling. Cell 1992; 70:401-410.

317. Rizzuto R, Brini M, Murgia M, Pozzan T. Microdomains with high Ca^{2+} close to IP_3-sensitive channels that are sensed by neighboring mitochondria. Science 1993; 262:744-747.

318. Roach TIA, Barton CH, Chatterje D, Blackwell JM. Macrophage activation: lipoarabinomannan from avirulent and virulent strains of *Mycobacterium tuberculosis* differentially induces the early genes c-*foc*, KC, JE, and tumor necrosis factor-α. J Immunol 1993; 150:1886-1896.

319. Roach TIA, Chatterje D, Blackwell JM. Induction of early-response genes KC and JE by mycobacterial lipoarabinomannans: regulation of KC expression in murine macrophages by *Lsh/Ity/Bcg* (Candidate *Nramp*). Infect Immun 1994; 62:1176-1184.

320. Robinson PJ, Millrain M, Antoniou J, Simpson E, Mellor AL. A glycosylphospholipid anchor is required for Qa-1-mediated T cell activation. Nature 1989; 342:85-87.

321. Rozakis-Adcock M, McGlade J, Mbamalu G, Pelicci G, Daly R, Li W, Batzer A, Thomas S, Brugge J, Pelicci PG, Schlessinger J, Pawson T. Association of the Shc and Grb-2/Sem-5 SH2-containing proteins is implied in activation of the Ras pathways by tyrosine kinases. Nature 1992; 360:689-692.

322. Saha AK, Dowling JN, LaMarco KL, Das S, Remaley AT, Olomu N, Pope MT, Glew RH. Properties of an acid phosphatase from *Legionella micdadei* which blocks superoxide anion production by human neutrophils. Archvs Biochem Biophys 1985; 243:150-160.

323. Saha AK, Dowling JN, Mukhopadhyay NK, Glew RH. Demonstration of two protein kinases in extracts of *Legionella micdadei*. J Gen Microbiol 1988; 134:1275-1281.

324. Saha AK, Dowling JN, Mukhopadhyay NK, Glew RH. *Legionella micdadei* protein kinase catalyzes phosphorylation of tubulin and phosphatidylinositol. J Bacteriol 1989; 171:5103-5110.

325. Saha AK, Dowling JN, Pasculle AW, Glew RH. *Legionella micdadei* phosphatase catalyzes the hydrolysis of phosphatidylinositol 4,5-bisphosphate in human neutrophils. Arch Biochem Biophys 1988; 265:94-104.

326. Sahney NN, Lambe BC, Summersgill JT, Miller RD. Inhibition of polymorphonuclear leukocyte function by *Legionella pneumophila* exoproducts. Microb Pathogen 1990; 9:117-125.

327. Samelson LE, Philips AF, Luong ET, Klausner RD. Association of the Fyn protein-tyrosine kinase with the T cell antigen receptor. Proc Natl Acad Sci USA 1990; 87:4358-4362.

328. Sando JJ, Maurer MC, Bolen EJ, Grisham CM. Role of cofactors in protein kinase C activation. Cell Signal 1992; 4:595-609.

329. Satoh T, Kaziro Y. Ras in signal transduction. Cancer Biol 1992; 3:169-177.

330. Satoh T, Nakafuku M, Kaziro Y. Function of Ras as a molecular switch in signal transduction. J Biol Chem 1992; 267:24149-24152.

331. Sawada S, Suzuki G, Kawase Y, Takaku F. Novel immunosuppres-

sive agent, FK506: in vitro effects on the cloned T cell activation. J Immunol 1987; 139:1797-1803.

332. Schlessinger LS. Macrophage phagocytosis of virulent but not attenuated strains of *Mycobacterium tuberculosis* is mediated by mannose receptors in addition to complement receptors. J Immunol 1993; 150:2920-2930.

333. Schlesinger LS, Bellinger-Kawahara CG, Payne NR, Horwitz MA. Phagocytosis of *Mycobacterium tuberculosis* is mediated by human monocyte complement receptors and component C3. J Immunol 1990; 144:2771-2780.

334. Schlessinger LS, Hull SR, Kaufman TM. Binding of terminal mannosyl units of lipoarabinomannan from a virulent strain of *Mycobacterium tuberculosis* to human macrophages. J Immunol 1994; 152:4070-4079.

335. Schmidt B, Konig S, Svergun D, Volkov V, Fischer G, Koch MH. Small-angle X-ray solution scattering study on the dimerization of the FKBP25mem from *Legionella pneumophila*. FEBS Lett 1995; 372:169-172.

336. Schmidt HHHW, Lohmann SM, Walter U. The nitric oxide and cGMP signal transduction system: regulation and mechanism of action. Biochim Biophys Acta 1993; 1178:153-175.

337. Schreiber S. Chemistry and biology of the immunophilins and their immunosuppressive ligands. Science 1991; 251:283-287.

338. Schuchat A, Lizano C, Broome CV, Swaminathan B, Kim C, Winn K. Outbreak of neonatal listeriosis associated with mineral oil. Pediat Infect Dis 1991; 10:183-189.

339. Schulz S, Yuen PST, Garbers DL. The expanding family of guanylyl cyclases. Trends Pharmacol Sci 1991; 12:116-120.

340. Schwan WR, Demuth A, Kuhn M, Goebel W. Phosphatidylinositol-specific phospholipase C from *Listeria monocytogenes* contributes to intracellular survival and growth of *Listeria innocua*. Infect Immun 1994; 62:4795-4803.

341. Settleman J, Narasimhan V, Foster LC, Weinberg RA. Molecular cloning of cDNAs encoding the GAP-associated protein p190: implication for a signaling pathway from Ras to nucleus. Cell 1992; 69:539-549.

342. Sheehan B, Kocks C, Dramsi S, Gouin E, Klarsfeld AD, Mengaud J, Cossart P. Molecular and genetic determinants of the *Listeria monocytogenes* infectious process. Cur Top Microbiol Immunol 1994; 192:187-216.

343. Shenoy-Scaria AM, Kwong J, Fujita T, Olszowy MW, Shaw AS, Lublin DM. Signal transduction through decay-accelerating factor: interaction of glycosyl-phosphatidylinositol anchor and protein tyrosine kinases p56[lck] and p59[fyn]. J Immunol 1992; 149:3535-3541.

344. Shou C, Farnsworth CL, Neel BG, Feig LA. Molecular cloning of cDNAs encoding a guanine-nucleotide-releasing factor for Ras p21. Nature 1992; 358:351-354.
345. Shyjan AW, deSauvage FJ, Gillet NA, Goeddel DV, Lowe DG. Molecular cloning of a retina-specific membrane guanylyl cyclase. Neuron 1992; 9:727-737.
346. Sieh M, Bolen JB, Weiss A. CD45 specifically modulates the binding of Lck to a phosphopeptide encompassing the negative regulatory tyrosine of Lck. EMBO J 1993; 12:315-312.
347. Siliciano JD, Morrow TA, Desidero SV. Itk, a T cell-specific tyrosine kinase gene inducible by Interleukin 2. Proc Natl Acad Sci USA 1992; 89:11194-11198.
348. Simons K, van Meer G. Lipid sorting in epithelial cells. Biochemistry 1988; 27:6197-6202.
349. Sokolovic Z, Riedel J, Goebel W. Surface associated, PrfA-regulated proteins of *Listeria monocytogenes* synthesized under stress conditions. Mol Microbiol 1993; 8:219-227.
350. Sorensen AL, Nagai S, Houen G, Andersen P, Andersen AB. Purification and characterization of a low-molecular mass T cell antigen secreted by *Mycobacterium tuberculosis*. Infect Immun 1995; 63:1710-1717.
351. Southwick FS, Purich DL. Inhibition of *Listeria* locomotion by mosquito oostatic factor, a natural oligoproline peptide uncoupler of profilin action. Infect Immun 1995; 63:182-190.
352. Speizer LA, Watson MJ, Brunton LL. Differential effects of omega-3 fish oils on protein kinase activities in vitro. Am J Physiol 1991; 261: E109-E114.
353. Spiegel AM. Heterotrimeric GTP-binding proteins: an expanding family of signal transducers. Med Res Rev 1992; 12:55-71.
354. Spiegel AM. G proteins in cellular control. Curr Op Cell Biol 1992; 4:203-211.
355. Spiegel AM, Backlund PS jr, Butrynski JE, Jones TLZ, Simonds WF. The G protein connection: molecular basis of membrane association. Trends Biochem Sci 1991; 16:338-341.
356. Stamler J. Redox signaling: nitrosylation and related target interactions of nitric oxide. Cell 1994; 78:931-936.
357. Stefanova I, Horejsi V, Ansotegui IJ, Knapp W, Stockinger H. GPI-anchored cell-surface molecules complexed to protein tyrosine kinases. Science 1991; 254:1016-1019.
358. Stokes RW, Speert DP. Lipoarabinomannan inhibits nonopsonic binding of *Mycobacterium tuberculosis* to murine macrophages. J Immunol 1995; 155:1361-1369.
359. Stone RL, Dixon JE. Protein-tyrosine phosphatases. J Biol Chem 1994; 269:31323-31326.

360. Stralfors P, Hiraga A, Cohen P. The protein phosphatases involved in cellular regulations: purification and characterization of the glycogen-bound form of protein phosphatase-1 from rabbit skeletal muscle. Eur J Biochem 1985; 149:295-303.

361. Streuli M, Krueger NX, Thai T, Tang M, Saito H. Distinct functional roles of the two intracellular phosphatase-like domains of the receptor-linked protein tyrosine phosphatases LCA and LAR. EMBO J 1990; 9:2399-2407.

362. Sturgill-Koszycki S, Schlessinger PH, Chakraborty P, Haddix PL, Collins HL, Fok AK, Allen RD, Gluck SL, Heuser J, Russel DG. Lack of acidification in *Mycobacterium* phagosomes produced by exclusion of the vesicular proton-ATPase. Science 1994; 263: 678-681.

363. Styblo K, Rouillon A. Estimated global incidence of smear-positive pulmonary tuberculosis: unreliability of officially reported figures on tuberculosis. Bull Int Union Tuberc 1981; 56:118-126.

364. Supattapone S, Worley PF, Baraban JM, Snyder SH. Solubilization, purification, and characterization of an inositol triphosphate receptor. J Biol Chem 1988; 263:1530-1534.

365. Szamel M, Resch K. T cell antigen receptor-induced signal-transduction pathways. Activation and function of protein kinases C in T lymphocytes. Eur J Biochem 1995; 228:1-5.

366. Szeto L, Shuman HA. The *Legionella pneumophila* major secretory protein, a protease, is not required for intracellular growth or cell killing. Infect Immun 1990; 58:2585-2592.

367. Tabouret M, DeRycke J, Dubray G. Analysis of surface proteins of *Listeria* in relation to species, serovar and pathogenicity. J Gen Microbiol 1992; 138:743-753.

368. Takai Y, Kishimoto A, Inoue M, Nishizuka Y. Studies on a cyclic nucleotide-independent protein kinase and its proenzyme in mammalian tissues. I. Purification and characterization of an active enzyme from bovine cerebellum. J Biol Chem 1977; 252:7603-7609.

369. Tamura S, Lynch KR, Larner J, Fox J, Yasui A, Kikuchi K, Suzuki Y, Tsuiki S. Molecular cloning of rat type 2C (IA) protein phosphatase mRNA. Proc Natl Acad Sci USA 1989; 86:1796-1800.

370. Tang P, Rosenshine I, Finlay BB. *Listeria monocytogenes*, an invasive bacterium, stimulate MAP kinase upon attachment to epithelial cells. Mol Biol Cell 1994; 5:455-464.

371. Taylor SS, Buechler JA, Yonemoto W. c-AMP-dependent protein kinase: framework for a diverse family of regulatory enzymes. Annu Rev Biochem 1990; 59:971-1005.

372. Thelen M, Dewald B, Baggiolini M. Neutrophil signal transduction and activation of the respiratory burst. Physiol Rev 1993; 73:797-821.

373. Thomas JE, Soriano P, Brugge JS. Phosphorylation of c-Src on tyrosine 527 by another protein tyrosine kinase. Science 1991; 254:568-571.

374. Thomas M. The leukocyte common antigen family. Annu Rev Immunol 1989; 7; 339-370.

375. Thomas PM, Samelson LE. The glycosylphosphatidylinositol anchored Thy-1 molecule interacts with the p60^{fyn} protein tyrosine kinase in T cells. J Biol Chem 1992; 267:12317-12322.

376. Thorpe DS, Niu S, Morkin E. Overexpression of dimeric guanylyl cyclase cores of an atrial natriuretic peptide receptor. Biochem Biophys Res Commun 1991; 180:538-544.

377. Thorpe TC, Miller RD. Extracellular enzymes of *Legionella pneumophila*. Infect Immun 1981; 33:632-635.

378. Tilney LG, Portnoy DA. Actin filaments and the growth, movement, and spread of the intracellular bacterial parasite, *Listeria monocytogenes*. J Cell Biol 1989; 109:1597-1608.

379. Tomioka H, Saito H. Phospholipids of *Mycobacterium intracellulare* inhibit T cell blastogenesis. Microbiology 1994; 140:829-837.

380. Tonks NK, Yang Q, Flint AJ, Gerbink MFBG, Franza BR jr, Hill DE, Sun H, Brady-Kalnay S. Protein tyrosine phosphatases: the problem of a growing family. Cold Spring Harbor Symp Quant Biol 1992; LVII: 87-94.

381. Trahey M, Wong G, Halenbeck R, Rubinfeld B, Martin GA, Ladner M, Long CM, Crosier WJ, Watt K, Koths K, McCormick F. Molecular cloning of two types of GAP complementary DNA from human placenta. Science 1988; 242:1697-1700.

382. Traylor TG, Sharma VS. Why NO? Biochemistry 1992; 31: 2847-2849.

383. Tsai M-H, Hall A, Wei FS, Stacey DW. Inhibition by phospholipids of the interaction between R-ras, Rho and their GTPase activating proteins. Mol Cell Biol 1989; 9:5260-5264.

384. Tsai M-H, Yu C-L, Stacey DW. A cytoplasmic protein inibits the GTPase activity of H-ras in a phospholipid-dependent manner. Science 1990; 250:982-985.

385. Tsygankov A, Bolen J. The Src family of tyrosine protein kinases in hemopoietic signal transduction. Stem Cells 1993; 11:371-380.

386. Ulrich A, Schlessinger J. Signal transduction by receptors with tyrosine kinase activity. Cell 1990; 61:203-212.

387. van der Geer P, Hunter T, Lindberg RA. Receptor protein-tyrosine kinases and their signal transduction pathways. Annu Rev Cell Biol 1994; 10:251-337.

388. Vazquez-Boland J-A, Kocks S, Dramsi S, Ohayon H, Geoffroy C, Mengoud J, Cossart P. Nucleotide sequence of the lecithinase operon of *Listeria monocytogenes* and possible role of lecithinase in

cell-to-cell spread. Infect Immun 1992; 60:219-230.

389. Velge P, Bottreau E, Kaeffer B, Yurdusev N, Pardon P, van Langendonck N. Protein kinase inhibitors block the entries of *Listeria monocytogenes* and *Listeria ivanovii* into epithelial cells. Microb Pathogen 1994; 17:37-50.

390. Venisse A, Fournie JJ, Puzo G. Mannosylated lipoarabinomannan interacts with phagocytes. Eur J Biochem 1995; 231:440-447.

391. Verma A, Hirsch DJ, Glatt CE, Ronnett GV, Synder SH. Carbon monoxide: a putative neural messenger. Science 1993; 259:381-384.

392. Vicente MF, Baquero F, Perez-Diaz JC. Cloning and expression of the *Listeria monocytogenes* haemolysin in *Escherichia coli*. FEMS Microbiol Lett 1985; 30:77-79.

393. Volpi M, Yassin R, Tao W, Molski TFP, Naccache PH, Sha'afi RI. Leukotriene B_4 mobilizes calcium without the breakdown of polyphosphoinositides and the production of phosphatidic acid in rabbit neutrophils. Proc Natl Acad Sci USA 1984; 81:5966-5969.

394. von Mollard GF, Stahl B, Li C, Sudhof TC, Jahn R. Rab proteins in regulated exocytosis. Trends Biochem Sci 1994; 19:164-168.

395. Wallace MR, Marchuk DA, Andersen LB, Letcher R, Pdeh HM, Saulino AM. Type 1 neurofibromatosis gene: identification of a large transcript disrupted in three NF1 patients. Science 1990; 249:181-186.

396. Walton KM, Dixon JE. Protein tyrosine phosphatases. Annu Rev Biochem 1993; 62:101-120.

397. Wei W, Mosteller RD, Sanyal P, Gonzales E, McKinney D, Dasgupta C, Li P, Liu B-X, Broek D. Identification of a mammalian gene structurally and functionally related to the CDC25 gene of *Saccharomyces cerevisiae*. Proc Natl Acad Sci USA 1992; 89:7100-7104.

398. Weiss A, Littman DR. Signal transduction by lymphocyte antigen receptor. Cell 1994; 76:263-274.

399. Wiesmuller L, Wittinghofer F. Signal transduction pathways involving Ras. Cell Signal 1994; 3:247-267.

400. Winn WC Jr, Myerowitz RL. The pathology of the *Legionella* pneumonias. Hum Pathol 1981; 12:401-422.

401. Wolf BA, Turk J, Sherman WR, McDaniel ML. Intracellular Ca^{2+} mobilization by arachidonic acid. Comparison with *myo*-inositol 1,4,5-triphosphate in isolated pancreatic islets. J Biol Chem 1986; 261:3501-3511.

402. Wolfe L, Francis SH, Corbin JD. Properties of a cGMP-dependent monomeric protein kinase from bovine aorta. J Biol Chem 1989; 264:4157-4162.

403. Woods GL, Washington JA II. Mycobacteria other than *Mycobacterium tuberculosis*: review of microbiologic and clinical aspects. Rev Infect Dis 1987; 9:275-294.

404. Xu S, Cooper A, Sturgill-Koszycki S, van Heyningen T, Chatterjee D, Orme I, Allen P, Russel DG. Intracellular trafficking in *Mycobacterium tuberculosis* and *Mycobacterium avium*-infected macrophages. J Immunol 1994; 153:2568-2578.

405. Yakura H. The role of protein tyrosine phosphatases in lymphocyte activation and differentiation. Crit Rev Immunol 1994; 14:311-336.

406. Yamamoto Y, Klein TW, Shinomiya H, Nakano M, Friedman H. Infection of macrophages with *Legionella pneumophila* induces phosphorylation of a 76-kilodalton protein. Infect Immun 1992; 60:3452-5.

407. Yamane HK, Fung BKK. Covalent modifications of G proteins. Annu Rev Pharmacol Toxicol 1993; 32:201-241.

408. Yoshida K, Asaoka Y, Nishizuka Y. Platelet activation by simultaneous actions of diacylglycerol and unsaturated fatty acids. Proc Natl Acad Sci USA 1992; 89:6443-6446.

409. Yuen PST, Garbers DL. Guanylyl cyclase-linked receptors. Annu Rev Neurosci 1992; 15:193-225.

410. Zeisel SH. Choline phospholipids: signal transduction and carcinogenesis. FASEB J 1993; 7:551-557.

411. Zhou M, Small SA, Kandel ER, Hawkins RD. Nitric oxide and carbon monoxide produce activity-dependent long-term synaptic enhancement in hippocampus. Science 1993; 260:1946-1950.

412. Zhukarev V, Ashton F, Sanger JM, Sanger JW, Shuman H. Organization and structure of actin filament bundles in *Listeria*-infected cells. Cell Motil Cytoskeleton 1995; 30:229-246.

INDEX

Page numbers in italics denote figures (f) or tables (t).